Gravity's Arc

Also by David Darling

Deep Time

Equations of Eternity

Soul Search

Zen Physics

The Extraterrestrial Encyclopedia

Life Everywhere: The Maverick Science of Astrobiology

The Complete Book of Spaceflight

The Universal Book of Astronomy

The Universal Book of Mathematics

Teleportation: The Impossible Leap

Gravity's Arc

The Story of Gravity, from Aristotle to Einstein and Beyond

DAVID DARLING

John Wiley & Sons, Inc.

To Jill, forever

Published by John Wiley & Sons, Inc., Hoboken, New Jersey
Published simultaneously in Canada

Design and composition by Navta Associates, Inc.

Library of Congress Cataloging-in-Publication Data:

Darling, David.
Gravity's arc : the story of gravity from Aristotle to Einstein and beyond / David Darling.
p. cm.
Includes bibliographical references and index.
ISBN-13 978-0-471-71989-2 (cloth)
ISBN-10 0-471-71989-7 (cloth)
1. Gravitation–History. 2. General relativity (Physics)–History. 3. Physicists–Biography. I. Title.
QC178.D373 2006
531'.14–dc22

2005030772

Printed in the United States of America

10 9 8 7 6 5 4 3 2 1

The roots of the Academy of twelve olive groves,
millennia later—the olives—Newton's laws;
today, the black olives, dark energy, in a hurry to arrive
beyond the barren terrain and isolated tranquility of spacetime.

—ANNA KANCHEVA, BULGARIAN POET

Contents

Acknowledgments

Among those have who been kind enough to provide technical advice and information are Stuart Anderson, Sasha Buchman, Bill Folkner, Joshua Gilder, Gary Page, Michael Salamon, Robert Soberman, and Robert Wald. Any errors that appear in the book arc, of course, my responsibility alone.

I'm once again very pleased to thank my outstanding editor, Stephen Power, and my production editor, Lisa Burstiner, at John Wiley & Sons, and my wonderful agent, Patricia Van der Leun.

Lastly and mostly, I'm grateful to my family for their love and support.

Prelude: The Weighting Game

We live in uneasy tension with gravity. Without gravity we wouldn't be here; it holds the atmosphere and the oceans to the earth, and it keeps the earth in orbit around the sun. It was the very reason the sun and Earth formed in the first place. Ultimately, gravity creates the conditions needed for life like ours to appear and survive. But it also creates enormous problems for those who live in its thrall.

Take our buildings, for example. The greatest of them soar elegantly in seeming defiance of the relentless force that tries to flatten them. But every one of them is a compromise, a visible, titanic struggle between aspiration and load. We want our buildings to float, to fly, to be filled with space and light. At the most basic level, however, gravity insists that you can't have a roof without support, so when the roof is too wide or weighty to be upheld by four simple walls, we need some other way to stop the edifice from crashing down around us.

The easiest answer to this problem of support is the column. Yet, however tastefully adorned, the column is the most pathetic of all supporting elements because it brutally exposes the architect's impotence. Those who built the Parthenon above Athens didn't want its interior to be so crowded with massive pillars, but they knew no other way to prevent the thing from collapsing. Only gradually, through trial and error, did the designers of buildings come to understand more intimately the play of forces and the use of shape and design to conduct gravity more subtly from rooftop to ground.

In the hands of the Arabs, the beam evolved to become the arch, which allows the supporting columns to be pushed further apart. The semicircular arch became the pointed arch, which adds strength and height. The arch was rotated in three dimensions to give the dome. Block-built pyramids and column-crammed, flat-beamed Greek temples gave way to spacious, sunlight-filled cathedrals with lofted ceilings, fan vaulting, and flying buttresses. The builders and designers of these finely tuned structures, the freemasons, carried in their heads a stock of ideas about how to control gravity, how to make stress flow to the outside of a building, that had grown from experience. The aqueduct at Segovia, the cathedral at Reims, and the Duomo of Florence reveal in their curved, ribbed, soaring forms the solution to one of gravity's challenges: how to build gracefully on a grand scale and at the same time manage the loading that gravity imposes.

Gravity is manifest not just in our buildings and in the natural landscape of mountains, caves, and rock bridges and cantilevers, but in our own bodies. It's literally in our bones, in their size and shape and arrangement. Millions of years of evolution have sculpted our bodies and internal processes, and those of other creatures, to survive and thrive in a one-*g* environment. When life emerged from the sea and became many-celled, it had to deal with the serious consequences of weight for the first time. Land species shifted their orientation with respect to gravity, or gained height, and so began to develop ways to cope with directional changes and to move structures and fluids against this load.

A fascinating instance of gravity playing a direct role in evolution concerns snakes. The orientation of a snake to the direction of gravity depends on habitat: tree snakes spend their days crawling up and down trees; land snakes spend most of their time in a horizontal position; sea snakes are neutrally buoyant. Researchers have found that of all snakes, the tree snake has its heart closest to its brain—an adaptation that allows it to get sufficient blood to its brain against gravity's pull. In humans, many of the most obvious signs of aging, such as sagging faces and bodies, rickety joints, and stooping, can be blamed on the relentless downward drag.

Everything around and inside us is adapted to existence in normal Earth gravity. Before the dawn of the space age, people wondered what would happen to human beings and other life-forms when thrust into a situation where the old rules didn't apply. In orbit, and while traveling at a steady speed through space, objects are weightless. While accelerating into space and reentering, spacecraft, and everything in them, experience several times Earth's gravity. Of prime concern was how well humans could tolerate such conditions, and for how long. The American physician John Stapp, as we'll discover, did some remarkable experiments on himself involving rocket-powered sleds that subjected him to several tens of *g*s for brief periods. Prospective astronauts were whirled around in centrifuges and other disorienting devices. Animals were sent on test flights to the edge of space. Eventually, humans spent many months at a time aboard space stations. The effects of microgravity on the human body became clear: muscle degeneration, space motion sickness, and bone demineralization.

Other animals, too, have trouble adapting to unearthly regimes. Spiders can't build proper webs without gravity as a guide. Zebra fish don't develop a normal vestibular system, essential for balance, if they grow up in a zero-gravity environment. Spaceflight studies in general show that gravity plays a crucial role in the development and health of vertebrates. Very-long-duration manned missions in zero-gravity will need some form of artificial gravity. Further down the road, other questions remain to be answered: Is the lower gravity on the moon and on Mars enough for people to live under for years at a stretch? Would children born on such worlds ever be able to come to Earth? Life will most likely look and move quite differently after several generations in space. Astrobiologists speculate, too, on what strange forms life might take if it evolves from scratch on a world where unusually low or high gravity prevails.

Each of us is a natural expert in gravity. We have to be in order to survive. The hunter who habitually misses his prey because he misjudges the reach of his spear or sling isn't going to bring home enough meat to feed himself or his family. Our ancestors, who spent much of their time in trees, swinging from branch to branch and leaping across chasms, would never have evolved further without an exquisite, built-in knowledge of how the deadly game of gravity must be played. To be able to process and coordinate a blizzard of visual and tactile signals on the fly, the primate brain grew disproportionately large. Gravity helped give us the very gray matter we now use in an effort to understand this mysterious, all-pervasive power of attraction.

Why do things fall? How quickly do they fall? What path do they follow? Why do some things, despite being apparently unsupported, *not* fall? The moon, the sun, the planets, and the stars are all "up there" in the sky, never deigning to come to Earth. What makes them different from a rock thrown skyward that quickly plummets to the ground? If gravity is a force, what impels it? What is the source of this strange, pervasive pull? It's taken several thousand years to answer these questions even partially. Within the past few years, we've found that gravity has a mysterious cosmic adversary. This so-called dark energy acts as a kind of antigravity, reversing gravity's trend to pull things together and threatening to stretch the cosmos into infinite oblivion. Physicists still struggle to fit gravity into an overall scheme of nature. The long journey to know the ultimate truth about gravity has not yet ended, but it began some two and a half millennia ago in ancient Greece.

1

No Laughing Matter

These, Gentlemen, are the opinions
upon which I base my facts.

—WINSTON CHURCHILL

Gravity is the mysterious attraction that holds us to the earth and, in general, draws together all things made of matter. Every schoolchild today is brought up with this idea in mind. Yet the notion of gravity as a force is fairly new; it dates back only to the seventeenth century, when Isaac Newton and his contemporaries began to rethink the way the world works. Before that time, gravity was seen in a very different light. It was assumed to be something an object *possessed*, a built-in property of its substance that decided how strong was the object's urge to fall. For almost two thousand years, this belief survived with barely a murmur of protest—an enduring brainchild of the first colossus of gravity.

A Meeting of Minds

Aristotle was born in 384 B.C. in Stagira, a Greek colony and seaport on the Thracian peninsula of Chalcidice, the part of Greece that looks like a three-fingered hand reaching into the northwest Aegean Sea. This is hilly country, tumbling down steeply to the blue salt waters, dense with low-lying fruit trees, bushes, and flowering shrubs, and interspersed here and there with outcrops of bright, bare rock. Aristotle's family was upper class, well connected, and intellectual; they were doctors by profession. His father, Nichomachus, served as court physician to King Amyntas III of Macedonia (the region to the north of Chalcidice), a connection that was instrumental in launching Aristotle's long association with the Macedonian royals, which would lead eventually to his tutoring Amyntas's celebrated grandson, the future Alexander the Great. Not much else is known of Aristotle's childhood. Both his parents died when he was young but probably not before Nichomachus had passed on to his son as much of his expertise as possible, a task he was duty bound to do by the Hippocratic Oath. In this way, Aristotle would have been exposed to some of the most advanced biological knowledge in the classical world, and he likely gained, early on, a deep curiosity about nature.

Although the Chalcidice peninsula is very much a part of modern-day Greece, it was considered rough and barbaric–an intellectual backwater–in those far-off days. All the big thinkers and seats of learning were congregated in a few key city-states, of which Athens, to the northwest, was preeminent. Just outside the city walls of the capital lay the Academy, the Harvard of the ancient world. At the head of the Academy was Plato, the foremost thinker of his age. At the age of eighteen, Aristotle was packed off by his guardian and mentor, Proxenus, to Athens to complete his education under this master of philosophy.

The Academy supposedly took its name from Hekademos, a mythical Attic hero at the time of the Trojan War, who, legend has it, planted twelve olive groves on land he owned about a mile from the center of Athens, using shoots from the sacred tree of Athena, the chief goddess of Greece, on the Acropolis. He then bequeathed the place for use as a public gymnasium (an athletic training ground) and shrine to Athena and other deities. Several hundred years later, in the sixth century B.C., Hippias, a tyrant of Athens, built a wall around the site and put up some statues and temples. Meanwhile, the statesman Kimon went so far as to have the course of the river Cephisus changed so that it would make the dry land of this popular park more fertile. Festivals were held there, as were athletic events in which runners would race between the various altars. Then, in about 387 B.C., Plato inherited a house nearby, together with a garden inside the grounds of the park. Here, within this pleasant, leafy retreat, he founded his Academy.

The Academy has often been described as the first university in the West–a fair enough description in the sense that it became a focus of intellectual energy, a place set aside from the workaday world to which keen minds could come to learn and discuss lofty ideas across a range of disciplines in seminars, informal talks, and meetings. Yet the only university-style lectures in the Academy were in mathematics. Plato is said to have had inscribed "Let no one who is not a geometer enter" above the door.

While that may be myth, given that the first reference to the inscription appears in a document written more than seven hundred years after Plato died, mathematics unquestionably loomed large in Plato's cosmic master plan. He was drawn to the subject because of its idealized abstractions, its transcendent purity, and the fact that it stood aloof from the material world, somehow above and beyond it. Natural philosophy–science, as we now call it–was anathema to him, an inferior and unworthy sort of knowledge. Mathematics in its most unadulterated form, Plato believed, could have nothing to do with the gross and imperfect goings-on in everyday life. Where it interfaced with reality at all, it must be well outside the flawed human realm–working at the most fundamental level, underpinning the very nature of things, and also, on the grandest of scales, encapsulating the structure of the universe as a whole. In such musings there's more than a whiff of intellectual snobbery: the aristocrat of knowledge, from a privileged family–his father's side claimed descent from the sea god Poseidon–waited upon by slaves, not wanting to deal with the sordid reality of commonplace data. But we can also glimpse an early attempt to devise a "theory of everything," a way of accounting for all of the most basic ingredients of nature within a unified mathematical framework.

The Fifth Element

At the heart of Plato's cosmological scheme lies the simplest and most perfect of three-dimensional geometric shapes, a point he drives home most emphatically in one of his later and best known dialogues, *Timaeus.* (The bulk of Plato's major writings take the form of contrived two-way conversations, often involving his teacher, the great Socrates.) In *Timaeus*, Plato talks about the five, and only five, possible regular solids–those with equivalent faces and with all lines and angles formed by those faces equal: the four-sided tetrahedron, the six-sided hexahedron or cube, the eight-sided octahedron, the twelve-sided dodecahedron, and the twenty-sided icosahedron. Today we call these shapes the Platonic solids because they first became widely known in medieval Europe through their exposure in *Timaeus.*

But Plato didn't discover them. Almost certainly, he learned of their existence during the ten years or so he spent in Sicily and southern Italy before setting up the Academy, probably from his close friend Archytas, a senior member of the Pythagorean school of thought. In fact, the bulk of Plato's knowledge and philosophy of mathematics was culled directly from the extraordinary Pythagorean sect.

Pythagoras, born in about 570 B.C. on the Ionian island of Samos, and his followers practiced a weird blend of mysticism and mathematics under the rubric "All is number." They lived by a litany of madcap rules, which forbade, for example, looking in a mirror by lamplight, eating beans, and putting one's shoe on the right foot first. They also held some eccentric beliefs (Pythagoras himself thought he was semidivine) as well as a few enlightened ones, including that men and women were equal–something virtually unheard of at the time. Crucially, they were the world's first pure mathematicians. As a good many pure mathematicians and theoretical physicists do today, they started from the premise that thought was a surer guide than the senses and intuition ranked above observation. From the Pythagoreans, Plato inherited his most unshakable conviction–that behind the world we see lies a more fundamental, eternal realm accessible only via the intellect. From them, too, he gained knowledge of the five regular solids. And although "Platonic solids" is a misnomer, Plato was genuinely original in how he interpreted the significance of these shapes. He linked them with the classical elements of earth, water, air, and fire, and, in so doing, formed a bridge between the mathematical and the material. In *Timaeus* he wrote, "To earth, then, let us assign the cubic form, for earth is the most immovable of the four and the most plastic of all bodies, and that which has the most stable bases must of necessity be of such a nature." Noting that the tetrahedron has the smallest volume for its surface area and the icosahedron the largest, Plato saw in these shapes the properties of dryness and wetness, respectively, and hence a correspondence with the elements fire and water. The octahedron, which rotates freely when held by two opposite corners, he regarded as a natural partner for air, the most mobile of the elemental quartet.

But there are *five* regular solids. To Plato, utterly convinced of the truth of his geometric worldview and of the unassailable power of intuition, this discrepancy between theory and observation could mean only one thing: there must be another element in addition to the four already known. There must be, in other words, a *quinta essentia* or *quintessence*, a "fifth essence," not familiar on Earth. Surely, he reasoned, this quintessence was the stuff of the heavens and its form the remaining regular solid–the dodecahedron. In support of his claim he noted that there were twelve sides on the dodecahedron and twelve signs of the zodiac–the constellations that the sun passes through in the course of a year. "God used this solid for the whole universe," he declared, "embroidering figures on it."

A twelve-sided cosmos? Dreamed up long before humanity fathomed the true nature of stars and galaxies and the immensity of space and time? It seems, on the face of it, just another quaint idea, surely long overtaken by events. But in October 2003, Jean-Pierre Luminet and his colleagues at the Paris Observatory published a paper in the journal *Nature* arguing, on the basis of data collected by the orbiting Wilkinson Microwave Anisotropy Probe (WMAP), that the universe does indeed take the shape of a dodecahedron.[19]

WMAP, launched in 2001, is designed to survey very precisely the so-called cosmic microwave background, the much-cooled glow of the vast explosion in which the universe began. The wavelength of this radiation is remarkably pure, but like a musical note, it has harmonics associated with it.

These harmonics reflect the shape of the object in which the waves are produced. In the case of a note, that object is the musical instrument upon which the note is played. In the case of the microwave background, the object is the universe itself. WMAP's measurements revealed that the second and third harmonics of the microwave radiation–the quadrupole and octupole–are weaker than expected. This weakness can be explained, according to the French team, if the universe is assumed to be finite and dodecahedron-shaped. Unfortunately, their model doesn't involve anything quite so simple as a giant Platonic solid. That is because an ordinary dodecahedron has a definite inside and outside and exists in "flat" space–the kind of space we're familiar with in everyday life and to which Euclid's geometry applies. What Luminet and his coworkers proposed is something called *dodecahedral space*, first described by their fellow countryman Henri Poincaré in the nineteenth century. Also known as a Poincaré manifold, this is a strange type of mathematical space that doesn't lend itself to being easily visualized. But the key point is that it has the same kind of symmetry as the dodecahedral cosmos that Plato had in mind.

Plato may have struck lucky on another point, too. It's easy to look at the classical elements–earth, water, air, and fire–and conclude that they have very little in common with the elements known to modern science: hydrogen, helium, carbon, iron, and the rest. But that's not really a fair comparison. It's true that the classical elements don't seem much like the elements of today's periodic table; they *do*, however, correspond very closely with what we now call the states of matter. Read earth for solid, water for liquid, air for gas, and fire for plasma (an ionized gas, often described as the fourth state of matter), and the ancients no longer seem so far off the mark. That leaves Plato's quintessence without a modern-day partner.

Nothing in twentieth-century science seems to correspond to this esoteric, celestial stuff. But then, without any warning, dark matter appears. At least four-fifths of all the mass in the universe, it turns out, consists of this invisible ingredient whose nature remains a subject of intense debate. Even more recently, astronomers have found evidence for another mysterious cosmic component quite unlike anything ever seen on Earth: dark energy. Both dark matter and dark energy have, for the purpose of various theories, been tagged "quintessence" by modern physicists who are mindful of Plato's seminal ideas. Both, as we'll see, have come to figure prominently in the recent story of gravity. But the person who first tackled gravity head-on wasn't Plato himself but his most stellar student, Aristotle.

A Man of Substance

When Aristotle came to the Academy in 367 B.C., Plato had already been at the helm for twenty years. For another two decades they worked alongside each other, first as mentor and pupil, then as colleagues, as Aristotle's stature grew and he took on more of a teaching role. During this time, however, the two men drifted apart philosophically and eventually came to hold radically different views on the nature of the world.

Plato was convinced that ultimate reality lay in ideas and what he called "forms," that is, perfect abstractions of things, which were knowable only through the trained mind. In his opinion, objects we see around us are no more than distortions of the truth–twisted reflections of some Platonic ideal. For example, a particular tree, which might have a branch or two missing, a gnarled trunk, or a carving of lovers' initials, is merely a flawed embodiment of the ideal form of a tree from which its existence derives. Outside of space and time, outside of materiality, is the one pure, transcendent Tree that allows us to identify the imperfect reflections of all particular trees around us. Only reason, guided by the proper use of logic, Plato insisted, makes the perception of such ideal forms possible.

Aristotle had quite a different metaphysical take on the world. For him, the wellspring of reality wasn't some strangely detached realm of intangible forms but what we see right in front of us–physical objects, the nitty-gritty of everyday life, knowable through immediate experience. He had a passion for collecting samples of anything and everything–alive, dead, or inanimate–and then pigeonholing them in what seemed to him a logical way. It's said that both Philip II, son of Amyntas III, with whom Aristotle went to school, and, later, Philip's son, Alexander the Great, showered Aristotle not only with funds for his studies but also with thousands of slaves to scour the land for new specimens. Even if these stories are a bit overblown, there's no denying that Aristotle was a tireless classifier and encyclopedist–the most outstanding accumulator and organizer of natural facts in the ancient world.

His philosophy was firmly rooted in the practical, the observable. Whereas Plato argued that individual things acquired their characteristics only by association with the forms that inhabited some ethereal never-never land, Aristotle held that the basis of reality existed in the actual world. First and foremost, there were individual things, living and nonliving, fashioned of what he called primary substance. These individuals made up species composed of a more universal, secondary substance. Species, in turn, fell naturally into different genera consisting of material still more generic than that of species. In contrast to Plato, who was very much a dualist, Aristotle saw matter and form as inseparable aspects of everything that existed. Matter was the raw building material of things–clay, the matter of which bricks were made, bricks, the matter of which walls were made, and so on. Matter was an object's *potential* to become an actual thing. Form was its reality, its shape, and the essence whereby it belonged to a certain class. A block of marble, for instance, had the potential to take on whatever form a sculptor chose, while a seed or embryo had the potential to grow into a particular living animal or plant.

Aristotle questioned, as he did most things, the basic elements of which all matter was composed. In the end, at least as far as our cosmic backyard was concerned, meaning everywhere closer than the moon, he went along with the four-element scheme of earth, water, air, and fire that had first been put forward by Empedocles about a century earlier. Each element, Aristotle argued, had a unique combination of primary qualities–hot or cold and wet or dry. The primary qualities of fire, for example, were hot and dry, while those of earth were cold and dry. As well as these traits, each element had an innate *motive power*, which tended to make it move in a particular direction, toward what Aristotle called its natural place. Two of the elements, earth and water, had the motive power of gravity, which tended to make them fall earthward. The other two elements, air and fire, had a completely different motive power, known as levity, or lightness, which acted in the opposite direction, radially away from Earth.

The important point, in Aristotle's view, was that levity wasn't just a feebler version of gravity. Something that has levity isn't less heavy; it's light in an absolute sense. If gravity is thought of as a tendency to sink, then levity is equivalent to buoyancy. Different elements sort themselves out by changing places, like an air bubble rising in water while the water fills in behind it; each element becomes the motive for the natural tendencies of the other elements to move. Given this way of looking at things, Aristotle was forced to conclude that the idea of empty space was nonsense. After all, a substance located in a void, not surrounded or motivated by any adjoining substance of differing tendencies, wouldn't have any reason to move. "Nature abhors a vacuum," Aristotle insisted, because it would make any kind of motion impossible.

Middle Earth

Along with Plato and most, but not all, other ancient philosophers, Aristotle never doubted that Earth sat at the exact center of a finite, spherical universe—a geocentric, human-focused cosmos that was simple and confined by today's standards. The inner region, which included Earth and all the surrounding space between Earth and the moon—the so-called sublunar domain—was composed of the four elements: earth, water, air, and fire. Because the element of earth possessed the most gravity, its tendency was always to sink to the middle, with water settling into a shell outside it. Air and fire both rose because of their levity, but fire, having the greater levity of the two, naturally drifted up to the outermost region.

Since the earthly sphere was imperfect, its elements didn't usually occur in their pure form but instead were combined into various substances with intermediate properties. Wood, for instance, was a mixture of all four elements, a fact that only became apparent when it burned. Only then did one see the flames of fire set free, the smoke and fumes of air, the water that oozed and bubbled out, and the earthy ash left behind when everything else had escaped and the residue had cooled. In the ideal case, which could never actually be realized, the elements of the sublunar realm would form a set of concentric shells: fire on the outside, followed by air and water, and, finally, at the center, a ball made of pure elemental earth.

The human world, as portrayed in the classical cosmos, was permanently disorganized and in a state of flux, with nothing quite where it was supposed to be. But beyond sublunar space lay the heavens, eternally perfect, composed exclusively of the fifth element, quintessence, or, as Aristotle and the Pythagoreans before him preferred to call it, *ether*. Each of the objects seen in the sky was fixed to its own transparent crystalline sphere: innermost the moon, then the sun, then the planets as far out as Saturn. More distant than Saturn was the heavenly sphere of the fixed stars, and beyond even that, the Deity who had created it all.

It was a universe divided, split into two completely distinct parts, each with its own makeup and code of behavior. There were the heavens–everything more remote than and including the moon–in which all motion was uniform, never ending, and perfectly circular about the center. Separately, there was Earth and the space immediately around it, in which imperfection and motion of a very different kind were the norm. If the heavens were the playground of astronomy, the sublunar domain was the province of an apparently very different science, physics, and the only place where, according to Aristotle, the twin properties of gravity and levity influenced how objects moved.

All earthly motion, said Aristotle, is either natural or "violent." Natural motion always happens in a dead-straight line along the radius of the universe (in other words, either directly toward or away from Earth's center) and eventually comes to a halt. This idea follows logically from the Aristotelian belief that everything has its natural place. An object made mostly of the element earth will try to get as close to the cosmic middle (exactly to the middle if it's 100 percent earth) as its constitution prescribes, and then it will stop. All natural motion involves things striving to get to where they're supposed to be by virtue of their elemental makeup, the urgency of their movement being dictated by the amount of gravity or levity they contain. From this follows one of Aristotle's key conclusions: the heavier an object (in other words, the more gravity it has), the faster it will fall. Something made of three-quarters earth and one-quarter air, for example, will drop more quickly than something made of half earth and half air. Reality seems to agree–at least at a passing glance. A light and airy thing such as a feather *does* fall more slowly than a heavy, "earthy" object such as a stone, while a levity-rich substance, such as smoke or a flame, actually rises.

Aristotle made another claim about natural motion. He said that how fast an object falls depends inversely on the density of the medium it's falling through; so, for example, the same body will fall twice as fast through a medium that's half as dense. Again, this seems to square pretty well with everyday experience. Drop a stone in air and it will plummet more quickly than the same stone released underwater. In putting forward these ideas, Aristotle became the first to propose *quantitative* rules about how things fall–rules, moreover, that were elegant, easy to grasp, and, superficially at least, credible.

A Symphony of Two Movements

Like the theories of so many scientists and philosophers, those of Aristotle were very much a product of the environment in which they were hatched. Two millennia ago, there were no cars or planes zipping around. There wasn't much in the way of moving machinery at all. What motion an inhabitant of ancient Greece saw around him tended to involve people and animals; it was motion that was *willed*, that had a purpose in taking the creature to someplace it would rather be, and therefore, as Aristotle perceived it, fulfilled the creature's nature. It took no great leap of imagination for him to account for the motion of things obviously *not* alive, such as a pebble dropped from the hand, by extending the concept of the nature of something to inanimate matter. In this way he came to formulate his laws of natural motion of objects in terms of the four elements purposefully seeking their rightful place in the order of things.

Aristotle also had definite ideas about motion that, in his view, was not natural. Throw a stone out into a lake; in Aristotle's opinion, that is an unnatural or what he called a violent movement. Here, he was using the term in its original sense–our word *violent* comes from the Latin *violentus*, which means simply "force." Such movement, he insisted, can happen only as long as there's a continuous pushing: the speed at which something moves (violently) being proportional to the strength of the push. Take away the push, he said, and the unnatural motion stops in an instant. Of course, Aristotle was no fool. He was well aware that projectiles carry on moving for some time after they've been thrown from a hand or shot from a bow or catapult. They don't immediately plunge vertically to Earth.

If Aristotle's idea about unnatural motion was right, there had to be some other kind of push that came into play once an object had been thrown. Two possible explanations suggested themselves. The first was that air, displaced from in front of a thrown object, somehow circulated around the object in a loop and ended up giving it a shove from behind. The second possibility was that the initial impulse, given to the object at the point of release, caused the entire column of air in front of the object to be pushed forward so that the moving shaft of air essentially drew the object forward along with it.

This second theory didn't look very promising, even in Aristotle's time, because in order to account for the continued sideways motion of the projectile, it relied on the continued sideways motion of another object, namely, the air. But this left a question mark over what caused the continued motion of the air and merely swapped one problem for another. On balance, the first explanation–the pushing vortex theory–seemed to Aristotle the better bet.

Intriguingly, in his discussion of these two rival explanations of motion, Aristotle came very near to a major breakthrough. He pointed out that in a void neither of the two theories would work, "so nothing could go on moving unless it were carried." But then he added, "Nor (if it did move) could a reason be assigned why the projectile should ever stop–for why here more than there? It must therefore either not move at all, or continue its movement without limit, unless some stronger force impedes it." This truth would be echoed twenty centuries later by another great physicist and be named after him. It is none other than Newton's first law.

As far as the heavens were concerned, Aristotle toed the party line, contending that all celestial motion was circular. Up there, beyond the moon, none of the complicated business of objects seeking their natural place ever occurred. Everything in the heavens was already where it was supposed to be, serenely pursuing the only kind of movement that, in a finite universe, could go on forever without changing–movement that simply cycled around and around.

Problems in Paradise

Providing you don't ask too many questions or scratch too deeply beneath the surface, the whole Aristotelian scheme can seem quite believable. Certainly, it was good enough for most people and credible enough to survive largely unchallenged for the best part of two thousand years. But a bit of nosing around soon reveals cracks in the theory.

Somewhere between here and the moon, for example, according to Aristotle, there's a switchover from the types of substance and motion found in the sublunar domain to those that prevail in the heavens. It can't be a gradual transition between the two realms because perfection and imperfection don't mix: you can't have earthly contamination gumming up the crystalline spheres. There can't even be an empty buffer zone, because Aristotle wouldn't tolerate a vacuum at any price. So there must be a sudden change–a concept that is a bit awkward to say the least.

On top of this, there's a problem with the simple circular motion that the planets were supposed to follow. It didn't exist, as every ancient astronomer who had made careful observations of the night sky knew. There's just no way to explain how the planets move in the sky by assuming that each goes around the Earth on a single round track. If you insist that circular motion is the only game in the heavens, you need wheels within wheels, spheres within spheres, moving this way and that around different axes to pull off the trick of planetary movements, even approximately.

The first person to take on the challenge of devising a workable system of celestial spheres was Eudoxus, the most talented mathematician and astronomer of his day, who built an observatory at Cnidus in the first half of the fourth century B.C. His ingenious arrangement of twenty-seven spheres–one for the fixed stars, three each for the sun and the moon, and four each for the five known planets, Venus, Mercury, Mars, Jupiter, and Saturn–was further elaborated by Callipus and then by Aristotle, who ended up calling upon fifty-six interconnected, gimbaling spheres to bring theory roughly into step with the dance of the heavens.

Aristotle's science of the sky, with some final polishing by Hipparchus and Claudius Ptolemy over the next couple of centuries, survived intact until the 1600s. What's more surprising is that his terrestrial science, including his views on falling objects and projectiles, did the same. Of course, it's always easy to pick holes in the ideas of an earlier age. A modern physicist has no trouble seeing that Aristotle's vortex theory of projectile motion, for instance, is a nonstarter. There's no way a vortex can impart a net positive motive force to an object; at best it can only cut down the drag. Aristotle couldn't have known this. He didn't have access to a wind tunnel or any other means of measuring the air flow around something.

But that's not the point. Aristotle probably wouldn't have used a wind tunnel even if he'd had one available. He just didn't do experiments. Yes, he was a great observer and classifier. As a biologist and, especially, a marine biologist, he was unparalleled: he dissected at least fifty species, including sea urchins and starfish, and his masterful description of the octopus's reproductive system remained unsurpassed until the nineteenth century. But when it came to checking the rules of physics that he'd devised, he just wasn't interested. It seems extraordinary. A heavier object falls faster than a lighter one, he insisted. Fair enough; that isn't a difficult proposition to test. You take two stones of the same material, one twice as heavy as the other. You stand at the top of a tall cliff and release both stones together. If you hear two separate cracks as the stones hit the rocks below at different times, the rule is shown to work (although you might have to do a series of more accurate experiments to check it thoroughly). If you hear a single crack, then something is obviously wrong. Why didn't Aristotle do a test like this, or have his assistants or slaves do it for him? And not just once, but many times, using different objects and locations?

Take another example. If Aristotle is to be believed, a projectile will carry on moving sideways (as well as up or down) as long as it keeps getting a push from the air that supposedly rushes in behind it to avoid the terrible vacuum. But at some point, he argues, while the object is in flight, this pushing will stop. The vortex effect runs out of steam; it gets tired, even exhausted. Then the object, no longer supported in its "violent" motion, does the decent, natural thing and plummets, because of its gravity, vertically to the ground. Again, this isn't a claim you simply have to take for granted. You can watch, sideways on from a distance, as someone throws a stone through the air, fires a rock from a catapult, or shoots an arrow from a bow. You can see with your own eyes if it's true that a projectile ends its flight with a straight-line drop out of the sky. Yet, as far as we know, Aristotle never troubled to put his ideas to such scrutiny.

In comparing Aristotle with Plato, it's easy and fair to conclude that, of the two, Aristotle was closer to being a scientist in the modern sense. That's because, unlike Plato, he did, at least, base his theories on observations of the real world. The trouble is, he didn't subsequently put his theories to the test, and that's a deeply unscientific approach. Aristotle and Plato both relied heavily on reason and logic to build their worldviews. Once they had their mental pictures of the universe in place–their grand scheme of things intellectualized–they weren't about to dirty their hands with experiments to see if reality happened to agree with them. If future observations didn't quite tie in with their prescribed philosophies, well, too bad for the observations.

Aristotle's Legacy

Plato died in 347 B.C., and by every measure of ability and achievement, Aristotle should have succeeded him as head of the Academy. Plato himself referred to him as "the intellect of his school." But Aristotle had fallen out of favor with the other seniors of the institution. It wasn't because he was a bit of a dandy, wearing rings on his fingers and cutting his hair fashionably short. It wasn't even his personality, which, if his enemies were to be believed, tended toward the arrogant and overbearing. Aristotle's political views were, it seems, what got him into trouble. In any event, leadership of the Academy passed to

Plato's nephew Speusippus, and Aristotle left for the court of his friend Hermeas, lord of the state of Atarneus in Mysia (a region in what is now Turkey). There he stayed for three years, marrying Pythias, the king's niece (later he married a second time, a woman named Herpyllis, who bore him a son, Nichomachus), before moving on to Mytilene on the Greek island of Lesbos after Hermeas was deposed by the Persians.

At the invitation of Philip of Macedonia, Aristotle became tutor to Philip's thirteen-year-old son, Alexander, a post he held for the next five years. When Philip died, Alexander succeeded to the throne and Aristotle returned to Athens, which he hadn't visited since the death of Plato. He found the Academy flourishing under a new head, Xenocrates. Platonism had become the dominant philosophy of Athens, but it was also in stagnation. So Aristotle set up his own school at a place called the Lyceum.

For the next thirteen years Aristotle devoted his energies to teaching and to composing his philosophical treatises. He is said to have given two kinds of lectures: more detailed discussions in the morning for an inner circle of advanced students and popular discourses in the evening for the general body of lovers of knowledge.

At the sudden death of Alexander in 323 B.C., the pro-Macedonian government in Athens was overthrown, and there was a general reaction against anything Macedonian. Charged, bogusly, with impiety, Aristotle was forced to flee to Chalcis in Euboea. Not long after, in 322 B.C., he came down with a stomach illness and died.

Aristotle's legacy was huge. He had surveyed the whole of human knowledge as it was known in the Mediterranean world in his time. More than any other thinker, he had determined the direction and content of Western intellectual history. He was the author of a philosophical and scientific system that through the centuries became the support and vehicle for both medieval Christian and Islamic scholastic thought. Until the end of the seventeenth century, Western culture was Aristotelian.

It isn't hard to see why. Aristotle's whole approach to studying nature fitted in neatly with Occidental theology. The idea that every organism was beautifully crafted for a particular function–its "final cause," as Aristotle called it–in the grand scheme of nature pointed to the conclusion that the world had been designed. Also, Aristotle was deeply interested in the concept of *nous*, or eternal intelligence, and this, too, made his work readily acceptable to the Church of the middle ages. Even his chauvinist views weren't out of line with male-dominated orthodox theology. "Full excellence," he insisted, could be realized only by the mature male adult of the upper class, not by women, children, barbarians (non-Greeks), or salaried "mechanics" (manual workers). Some of his other silly ideas, such as that women had fewer teeth than men and that a baby's sex was determined by the wind's direction at the time of its birth, could be safely swept under the philosophical carpet. Backed by the Church, Aristotle's worldview was secure. His laws of motion and his ideas about gravity would stand–as long as no one looked at them too closely.

2

The Path of Dissent

The difficulty lies, not in the new ideas, but in escaping the old ones, which ramify, for those brought up as most of us have been, into every corner of our minds.

—JOHN MAYNARD KEYNES

Let something tumble from a great height and it makes a bigger thud than if dropped from a lesser height. That doesn't seem like an earth-shattering observation, but it was one of the arguments used by the first person to question Aristotle's theory of how objects fall. That person was a man called Strato.

Strato was born in the city of Lampsacus in Mysia, the region in Asia Minor where Aristotle had spent three years. In fact, Strato's career was strangely parallel to Aristotle's. Just as Aristotle had gone to Mysia after his time at the Academy to tutor Alexander, the king of Macedonia's son, so Strato went to tutor the future Ptolemy II after attending Aristotle's Lyceum. The boy was the son of Ptolemy I, who had been put in charge of Egypt by Alexander, his boyhood friend, following Alexander's conquest of most of the known world. Upon his return to Athens, Strato rejoined the Lyceum and in about 287 B.C. became its third director following the death of Theophrastus, who had taken over the post from Aristotle.

It's often said that, despite his genius, Aristotle held up the progress of science for many centuries. He did nothing of the sort. The fact that not enough people questioned his teachings sooner and more decisively wasn't his fault. What's more, his Lyceum became, for a while, an important focus of scientific observation and experimentation–a center where breakthroughs were made that could have drastically cut the time needed to transform our view of the cosmos. Strato himself, the "Physicist" as he became known for his inquiring approach to nature, has been called the father of experimental science, though Galileo probably deserves the title more.

Like Aristotle, Strato believed in the importance of observing natural phenomena. In the study of motion, however, he paid much closer attention to detail than did Aristotle. As we've seen, Strato noted that an object falling from a greater height lands with a bigger impact. He also pointed out that rainwater pouring off a corner of a roof is clearly moving faster when it hits the ground than when it leaves the roof. This is because a continuous stream can be seen to break into drops, which then spread further apart as they head earthward. These two observations exposed a major flaw in Aristotle's thinking.

Aristotle taught that a falling body reaches a constant speed of descent–what today we'd call its terminal velocity–almost immediately and then falls most of the rest of the way at a steady rate. To be fair, he also mentioned that as the body approaches more closely its own element, its weight increases and there is some acceleration; however, he was brief and vague on this point and certainly not quantitative.

Strato's observations clearly showed that falling bodies accelerate gradually over an extended period of time. Contrary to what Aristotle said, the impact of a body when it hits the ground doesn't just depend on its weight; it has to do with the height from which it falls as well.

Strato also took Aristotle to task on other issues. Objects aren't just heavy or light, he protested, but have a range of weights. He also denied the existence of levity–the supposed internal property of light objects that pushes them upward. Air and fire rise, Strato said, not because of levity, but because they're displaced by heavier bodies; in other words, they're pulled down less than the heavier objects around them. Had these lines of inquiry been taken further at the Lyceum, we might have been saved a thousand years or more of stagnation in physics. But after Strato died, the Lyceum drifted away from science and instead channeled its efforts into literary criticism.

Decline and Fall

The Alexandrine period, in general, was an amazingly fertile one for scientific and technological insights. During that time, Eratosthenes figured out the distance around the Earth to within less than 1 percent of its true value; Hero of Alexandria built a miniature working steam turbine; and Aristarchus gave an accurate picture of the solar system, with the sun, not Earth, at its center. But none of these advances was taken any further and most were eventually forgotten.

Part of the blame for that lost opportunity rests with the slave economy of the ancient world, which discouraged any association of science and technology. With a few exceptions, such as building better machines for war, rulers weren't interested in how scientific discoveries might be applied. Philosophical and scientific speculation was seen as a pointless intellectual pastime for the wealthy, whereas philosophers and mathematicians looked with contempt at men of practical affairs. Euclid, the great geometrician, when asked by a pupil what he could expect to earn from studying geometry, ordered a slave to give him a few coins, "since he must make a gain out of what he learns."

With cheap slave labor in plentiful supply, there wasn't any incentive to develop labor-saving technology. The market for refined products was restricted to a fairly small group of well-to-do folk, so the question of mass production never arose. With no economic incentive to turn pure science and math into technology, there was no tradition of inventing tools and instrumentation–telescopes, microscopes, measuring devices–that might furnish new data to test theories and impel the progress of science.

Greek civilization fell into decay, followed by European culture as a whole. All the brilliant innovations and ideas that had blazed into life around the Mediterranean rim were snuffed out or hidden away. As Europe plunged into the Dark Ages, the only body of knowledge that was tolerated was the sterile, unchanging dogma approved by the Church.

It's hard to imagine the utter rigidity of life and thought in those barren times–a thousand-year intellectual wilderness stretching from a few centuries A.D. to the dawn of the Renaissance. Society was rooted in feudalism, a system that locked everyone and his or her offspring, from peasant to king, into a strict hierarchy that offered no prospect of upward mobility or escape. This frozen pyramidal structure, and the unchanging character of the feudal mode of production that underpinned it, was echoed in the fixed teachings of the Church. Nothing less than unquestioning obedience, based on the official interpretation of Holy Scripture, was demanded by Rome. Blind faith supplanted reason, and science, with its pagan heritage, was looked upon with the deepest suspicion. Woe betide anyone who stepped out of line. The die was cast when one of the last of the Greek mathematicians, Hypatia, was stoned to death by a mob led by a monk.

Because of the Church's monopoly of culture, all intellectual life was channeled through it. At the medieval universities, such as Paris and Oxford, where everything was taught in Latin, the curriculum was dominated by grammar, logic, rhetoric, arithmetic, astronomy–Earth-centered and of the crystalline-spheres variety–and music. The high points were philosophy and theology, which were closely tied. For centuries, philosophy was seen as the handmaiden of theology. Science was stripped down to a bare minimum, and there wasn't the slightest interest in research and experiment. To seek out new knowledge was to run the risk of straying from the narrow prescribed path of virtue.

Philosophy was reduced to an impoverished form of Platonic idealism, later replaced by a completely ossified and one-sided reading of Aristotle. In the early period, St. Augustine based his arguments attacking the pagan opponents of Christianity on neo-Platonism. Much later, in the thirteenth century, the writings of St. Thomas Aquinas presented a version of Aristotelian philosophy trimmed and tailored to suit the needs of the Church. The theory of the cosmos, according to Aquinas, played down the materialist elements of Aristotle and stressed some of his weaker arguments, including the idea that the ultimate source of rotation of the stars and planets is the *primum mobile*, the "first movable" or outermost celestial sphere, which is itself unmoved. In Christian thought the answer to what lay beyond the primum mobile was "God," who existed outside space and time. But Aristotle never mentioned God, saying simply, in his *De Caelo* (On the Heavens), "Outside the heaven there is neither place nor void nor time. Hence whatever is there is of such a kind as not to occupy space, nor does time affect it."

The picture of the physical universe thus adopted by the Church was a reflection of the medieval world that its spiritual police wanted to see, and with the same kind of static, unchanging character. The hierarchy of society was faithfully reproduced in the tiered nature of the cosmos. Just as there was the pope, archbishops, bishops, and so on, down to the minor nobles who owned the land, so there was a celestial hierarchy of angels. Each of these heavenly beings had a definite job to do in running the universe and was attached according to rank to the planetary spheres, which kept each one in his or her proper place and state of motion. The lowest order of angels, bound to the sphere of the moon, naturally had the most to do with human affairs, which went on in the sublunar realm immediately below. In general there was a cosmic order, a social order, and an order inside the human body, all representing states to which nature tended to return when it was disturbed. Aristotle may have been a pagan, but that could be conveniently ignored. His spheres-within-spheres cosmology and his notion that everything had its natural place were, when suitably sanitized, perfect for assimilation by the Church.

A Change of Course

For centuries, the Islamic universities in Spain, especially the one at Cordoba, were the only vibrant centers of learning in Europe, except for Ireland, which, because of its remoteness, remained outside the mainstream. The Arabs made tremendous strides in mathematics, astronomy, geography, medicine, optics, and chemistry, as well as in technology. But it wasn't until the twelfth century that this knowledge began to percolate through to the West. Its effect then was to stir an undercurrent of dissent–the first murmurings of revolt against the stranglehold of a ruthless ecclesiastical mafia.

An early rebel was Roger Bacon, born in about 1214, who studied and taught at Oxford and became a Franciscan friar. It was impossible to get a decent education in those days without also becoming deeply immersed in Church dogma. But Bacon had the blood of a maverick; he was drawn to new ideas, to challenging the status quo, and, most dangerous of all, given the spirit of the times, to the experimental study of nature. His writings sometimes sound like wizardry, not surprising given that science still hadn't detached itself from alchemy and astrology. But among the mumbo-jumbo are bold ideas for calendar reform, early concepts for aircraft and submarines–long predating similar ones by Leonardo–and what sounds like a description of rockets:

> We can, with saltpeter and other substances, compose artificially a fire that can be launched over long distances. . . . By only using a very small quantity of this material much light can be created accompanied by a horrible fracas. It is possible with it to destroy a town or an army. . . . In order to produce this artificial lightning and thunder it is necessary to take saltpeter, sulfur, and Luru Vopo Vir Can Utriet.

The last five mysterious words make up an anagram that, when solved, gives the proportion of powdered charcoal needed to complete the explosive mixture. This explosive is none other than gunpowder or black powder–the key to a new generation of weapons and, indirectly, to a breakthrough in our understanding of gravity.

Bacon didn't invent gunpowder–the Chinese had used it in their firecrackers and "fire arrows" as early as the ninth century–but he was one of the first Westerners to learn its composition and pass along the recipe, albeit cryptically. At about the same time, the first weapons in Europe to exploit gunpowder began to appear. Prominent among them was the cannon. Again, the history of this development is lost in obscurity. Certainly, the Arabs had been using a variety of cannon before the device migrated north and west. But rightly or wrongly, credit for the smooth-bore brass cannon, which found its way onto the battlefields of Europe in about the thirteenth century, is often given to a German monk named Berthold Schwarz. As a weapon of terror, the cannon, belching fire, ear-splitting noise, and mayhem, was unrivaled. Its effectiveness in war was severely compromised, however, because no one could accurately tell where the shot would land after it had been blasted out of the tube. It might just as easily blow up your own troops as the enemy's.

The same problem bedeviled all instruments that fired projectiles. Suppose you were an English archer with a longbow. How far should you draw back the cord and how much should you tilt the bow so that your arrow would land smack among the French infantry marching toward you a field away? The longbow was one of the most formidable weapons of its time. An experienced archer could fire a dozen or more arrows a minute with a range of up to two hundred yards. But there were no magic formulae or precisely worded instructions to tell you what to do to hit a specific target at such-and-such a distance. You had to learn the hard way, by years of trial and error. During the heyday of longbow archery, beginning in late-twelfth-century England, every able-bodied man had to practice the skill on Sundays and holidays. The king's archers practiced six to eight hours a day, seven days a week. They came to know by long experience, and by tips passed on from one generation to the next, how to aim their bows to maximum effect.

Hands-on familiarity, too, was the only way to tame the trebuchet, the ultimate hurling machine. The simple trebuchet is just a big lever with a heavy weight at its short end and a projectile basket at the long end. The long end is pulled back with winch-drawn ropes. When the lever is released, even a heavy missile at the other end of the lever can reach a very high velocity. An interesting improvement on this simple catapult design was to attach a rope to the long end of the lever. The rope would then swing up as the lever was released and bring the projectile with it. At the appropriate point on the swing trajectory the rope would be cut to release the missile. An awesome siege weapon, the trebuchet was able to heave large objects into and over the walls of medieval fortresses. (Tests with a modern replica have shown it can throw a small car almost the length of a football field.) There was no methodical way of aiming the thing, however, and no dials you could set or tables you could look up to determine the range. It was all up to the skill of the weaponeer, aided by rules of thumb–secrets of the trade–handed down from master to apprentice.

There came a time, however, when this approach simply wasn't good enough. As weapons became more powerful and the outcome of a battle hinged on the accuracy with which missiles could be flung at the enemy, strategists needed a more scientific way of aiming. They wanted, if possible, to put ballistics on a quantitative footing, to be able to accurately foretell, for instance, the angle with which a device must be pointed to reach its target.

The best that Church-approved medieval science could do was to parrot Aristotle's ideas about the way things moved through the air. Aristotle had suggested that the shot from a cannon follows a path made up of two straight lines joined by the arc of a circle. The first straight line, inclined to the ground at the angle the shot is initially fired, marks the phase of violent motion when the projectile is supposedly being kept moving sideways by air rushing in behind it. The curved part of the path represents the so-called mixed motion that occurs when the pushing effect of the air begins to tail off. Finally, the second straight segment of the trajectory is the vertical drop to the ground as the projectile reverts to natural motion.

Anyone used to watching the paths of thrown objects would have been able to tell right away that there was something seriously wrong with this description. Archers and those manning devices like the trebuchet, with years of experience, couldn't have failed to notice that projectiles, especially those fired at fairly shallow angles to the ground, didn't end their flight abruptly by dropping straight down out of the sky.

The trouble was that the people preaching the theory from their ivory towers and those dealing with real-world situations were breeds apart; they never interacted. A peasant archer who knew something about the true flight of an arrow wouldn't have been concerned with high-brow theoretical arguments. On the other hand, until the beginning of the Renaissance, most natural philosophers considered it beneath their dignity–even sacrilegious–to put received wisdom to the test.

Stuttering Progress

It's a measure of how much the social climate had changed by the early sixteenth century that the Italian principality of Verona hired a mathematician as a ballistics consultant. Niccolò Tartaglia was asked to solve a simple-sounding problem: At what angle must a gun be held to fire the farthest?

Tartaglia wasn't his family name. He was born in 1499 in Brescia, a town in what was then the Republic of Venice, as

Niccolò Fontana. His father, a postal courier, died a few years later, leaving the family impoverished. Further disaster struck when the French sacked Brescia in 1512 and Niccolò and his mother sought refuge in the cathedral. The invading soldiers broke into the church and Niccolò was brutally assaulted, his skull split and his jaw and palate slashed through by a saber. Although he survived, his speech was so badly affected that he acquired the nickname Tartaglia, which in Italian means "stammerer," a name he later officially adopted. In time, he also grew a long beard to hide his horrific scars.

When he was about fourteen, Niccolò was sent to study under one Maestro Francesco, but his mother soon ran out of money to pay for the lessons, and, as Tartaglia wrote in his autobiography, "I never returned to a tutor, but continued to labor by myself over the works of dead men, accompanied only by the daughter of poverty that is called industry." Lacking the means even to buy paper, he was obliged to use tombstones as slates. He eventually mastered enough mathematics so that sometime between 1516 and 1518 he was able to move to Verona and take a job teaching the subject. Subsequently, he got married and began to raise a family, but he always struggled financially. In 1534 he moved to Venice and, on and off, continued to eke out a poor living as a mathematics teacher there. What brought him fame was the work he did in his spare time.

Tartaglia was instrumental in solving the kind of problems known as cubic equations—equations in which the highest power of the unknown, usually denoted by *x*, is 3. The story of cubics could fill a book in itself. The first mathematician known to have solved this type of equation using algebra was Scipione del Ferro, but, unfortunately, he didn't tell anyone until, on his deathbed, he passed on some of his secrets to one of his students, a man called Fior. Although only a mediocre mathematician himself, Fior then began bragging that he could solve cubics.

Meanwhile, Tartaglia had devised his own methods for tackling these equations, and so, in 1535, a debate was arranged between the two men to settle the question of who knew his subject better. Because Tartaglia had discovered how to solve all cubics, whereas Fior had been taught to unravel only some, the "Stammerer" easily came out on top. But the matter didn't rest there. Another Italian mathematician, Girolamo Cardano, heard of the contest and tried to get Tartaglia to divulge his cubic-solving techniques, which hadn't yet been published. Eventually, Tartaglia agreed, but only after Cardano swore to him that he would never reveal Tartaglia's discoveries to anyone. What followed was one of the most bitter disputes in the history of mathematics.

Based on Tartaglia's formula, Cardano and his assistant, Ferrari, made rapid progress not only in finding proofs of all cases of the cubic but, even more impressively, solving the quartic equation, which involves the fourth power of x. Tartaglia still hadn't gone public with his formula, despite the fact that, by then, it was well known that such a method existed. So a major argument was guaranteed when, in 1545, Cardano published his own solutions to the cubic and quartic equations, together with all of the additional work he'd done on Tartaglia's formula. To be fair, Cardano did give full credit in his text to both del Ferro and Tartaglia for their contributions.

Nevertheless, Tartaglia was furious when he found out that, as he saw it, Cardano had gone back on his word. In an effort to regain the upper hand, Tartaglia agreed to debate Ferrari. But although Tartaglia was vastly experienced in such mathematical showdowns and expected to win, it became clear by the end of the first day that things weren't going his way. Ferrari turned out to have a better overall grasp of cubic and quartic equations, and Tartaglia, realizing this, decided to cut and run from Milan, the site of the contest, before his shortcomings were further exposed. Victory was thus handed to his opponent, even though Tartaglia had done the more original work.

Tartaglia was a pioneer in other ways, too. He was the first Italian translator and publisher of Euclid's *Elements* in 1543. Early in his career, he began a fascination with the mathematics of warfare that led to some important breakthroughs in the understanding of projectile motion. For example, he was able to figure out the answer to the problem put to him by the military men of Verona: to maximize a gun's range, he found, you should tilt the barrel at an angle of 45 degrees. But Tartaglia went beyond mere calculation. He lived at the dawn of a new age in which science was becoming more practical and experimental. Complicating factors were at work that affected the performance of a gun or a cannon–factors such as air resistance–that were difficult to allow for theoretically. So, for every kind of gun he could lay his hands on, Tartaglia ran tests to see how range varied with elevation. The resulting firing tables, the first in history, appeared in his *New Science of Ballistics*, published in 1537, and proved invaluable for training future ballisticians.

Tartaglia measured elevation by using a quadrant. This was pretty much the same way that astronomers took the height of celestial objects above the horizon. But Tartaglia developed a special form of this device, called the gunner's quadrant, as an aid in aiming. This consisted of two wooden arms at right angles, between which was an arc marked with twelve divisions, known as points, with a plumb line attached at the angle. To use the quadrant, a gunner inserted the longer of its two arms in the cannon's muzzle and read the point at which the plumb line crossed the arc. When the gun was aimed vertically, the plumb line crossed the quadrant's arc at the twelfth division, known as "point twelve." When the gun was horizontal, the plumb line crossed the arc at zero.

Back in the sixteenth century, most people still used Roman numerals, and, strange as it seems, they hadn't yet fully grasped the concept of zero. So instead of calling this reading "point zero," they referred to it by an expression that remains familiar to this day–point blank.

One of Tartaglia's biggest breakthroughs was to realize that Aristotle's ideas about the shape of a projectile's path were wrong. The trajectory didn't start out as an inclined straight line nor end as a vertical line. On the contrary, said Tartaglia, the path of a projectile was curved all the way along, from start to finish. One of the illustrations from his *New Science* shows a gracefully looping arc of cannon shot while, in the background, Aristotle and others look on. He also claimed, correctly, that the greater the muzzle velocity, the flatter the projectile's path would be. He still subscribed, however, to some ancient and mistaken views about what happened to an object while it was in flight. Here he is explaining how the speed of a cannonball makes it lighter, causing it to float in the air until friction slows it down:

> The great speed is the true cause that holds the ball's motion to straightness, if that is possible. And similarly, the loss of speed in the air is the true cause that makes it tend and decline in its motion curvedly toward the earth, and the more it loses its speed there, the greater becomes its declining curvature. And all this happens because for any heavy body driven violently through the air, the faster it goes, the less heavy it is in that motion, and heretofore the straighter it goes through the air because the air more easily sustains a body the lighter it is.

Time would show that this picture of events was little better than fantasy. But no matter. With his curved trajectories, Tartaglia had already broken the mold of Aristotelian physics and shown the way forward for the science of ballistics with his experimental approach.

To say that a path is curved raises the question of the exact shape of the curve. One of the greatest investigative minds of all time would supply the answer, as we'll see in the next chapter. But his answer, though crucially important to science, didn't take account of the effects of air. For heavy, fairly slow-moving projectiles, this wasn't such an issue. What we now call drag, or air resistance, only gets significant as an object's velocity becomes high. As firearms continued to evolve and muzzle velocities rose, however, it was vital for gunners to be able to factor in the complicating effects of air in order to fire their weapons accurately.

On Target

Just as Tartaglia revolutionized ballistics in the 1500s, the English engineer Benjamin Robins triggered another seismic shift in the subject two centuries later with his invention of the ballistic pendulum. This was basically a massive iron plate bolted to the back of a block of wood that was suspended by strings. When a bullet slammed into the block, it caused it to move back and up, indicating the amount of energy that had been transferred. By measuring the distance traveled in an arc by the block and knowing the masses of the block and the bullet, it was easy to figure out the velocity at which the bullet had been traveling.

Robins used the ballistic pendulum to determine the velocities of shots fired at different ranges. By seeing how much the bullets slowed with range, he was able to calculate the extent to which air affects a projectile. According to his experiments, the drag caused by air had an influence on the projectiles he fired that was eighty-five times greater than that of gravity. Although these numbers seemed beyond belief at the time, they were proven time and again to be true. Robins had not only provided a more accurate way to predict the path of a projectile, but he had also shown the vital importance of drag to the field of ballistics.

Even with these new insights, however, producing gunnery tables by hand remained a long, slow, error-prone business. One of the biggest motivations for developing computers, first mechanical and then electronic, was to do artillery calculations much faster and more reliably. Charles Babbage's famous Difference Engine, a prototype computer built in 1822, was first put to use figuring out gunnery tables for the Royal Artillery. Before that, the military relied on pen, paper, and the human brain–not the most efficient of number crunchers.

The hazards of relying on dodgy gunnery tables were thrown into sharp relief during the Battle of Waterloo, when it became evident that Napoleon's cannons were a lot more accurate than those of the British army. The duke of Wellington left the field victorious but determined to sort out the problem, and, on his return to London, put the question to Captain Wilberforce Briggs of the Ordnance Survey. A less free-minded thinker might have taken the easy route and co-opted a group of bright mathematicians to produce updated tables by the tried-and-tested method.

Briggs, however, chose an entirely novel approach. He was aware that great players in the sports he knew–cricket, soccer, and golf–had an ability to *intuit* where the ball would fly in response to a vast number of subliminal factors. What was needed, he reasoned, wasn't another set of dubious tables, but a new breed of battery commanders who could *feel* where the shot would fall.

Taking his inspiration from the medieval archery statutes, which required the common man to practice shooting and gave the English military a pool of the best archers in Europe to draw from, he devised a game based on the French *flechettes* (a game that is similar to lawn darts). But whereas the French game used small weighted arrows and a target like a roundel in archery, Briggs's version involved an absurdly complicated target board, divided into segments numbered 1 to 20, with smaller compartments that merited higher scores.

By throwing weighted, finned darts at the target, Briggs surmised, artillerymen could hone their skills to twofold advantage. First, they would acquire a deeper understanding of the flight of a stabilized projectile. And second, they would develop mathematical skills, which would let them communicate their intuitions to the gunnery crews used to working in degrees of elevation and measures of powder. The game was subsequently installed in public houses throughout the land, and, with Wellington's backing, a bill was pushed through Parliament to make the practice of the game compulsory on Friday and Saturday evenings.

There are still some in Britain who consider this practice more or less essential and who, as a result, are able, while in a mild state of intoxication, to calculate quite elaborate sums, especially if they involve subtraction from 501. But sadly, English Darts, as it has become known in other countries keen to distance themselves from the invention, did nothing to improve the accuracy of eighteenth-century British guns. Having a feel for the flight of little feathered spikes, it turns out, doesn't help much with mastering the vagaries of field artillery. Unable to cope with the shame of his failure, Captain

Briggs left the military and retired to keep bees on the Sussex coast.

Dangerous Questions

It was one thing to tinker with science to improve the odds of winning a battle, but quite another to challenge the accepted order of the heavens–and thus, by association, the authority of Rome. Yet that's exactly what some medieval philosophers began to do as early as the fourteenth century. This questioning of nature in an effort to reconcile Aristotle's teachings with those of the Church started in the venerable universities of Paris and Oxford. Often it took the form of asking whether there might be inhabited worlds other than Earth. Inevitably, this debate about the "plurality of worlds" led to a theological controversy about whether God made Earth and its life uniquely, even though He surely had the option of creating other habitable places, or whether God encouraged life to bloom at every opportunity. The suggestion that Earth might not be the only abode of life raised the possibility that it might not be unique in other ways–in particular, that it might not occupy a privileged position at the center of the cosmos. That was a risky speculation indeed.

Jean Buridan, rector of the University of Paris in the first half of the fourteenth century, was early on the pluralist scene, arguing that "[I]t must be realized that while another world than this is not possible naturally, this is possible simply speaking, since we hold from faith that just as God made this world, so he could have made several worlds." His contemporary, William of Occam, who joined the Franciscan order and studied at Oxford, went a step further by pointing out that if there were multiple Earths, Aristotle's doctrine of natural place would presumably apply to each of them–the elements of each returning to their natural place within their own world, without any intervention by God. Although, later on, Occam fell into line with an orthodox reading of Scripture, which entertains one and only one Creation, and shied away from allowing that there might be other Earths, he was instrumental in encouraging a new generation of more liberal thinkers, such as his pupil Nicholas of Oresme.

Master at the University of Paris and bishop of Lisieux, Oresme taught that Aristotle's doctrine of natural place was valid, providing only that heavy bodies were located more centrally than light ones. Since there could be many centers, there could, in principle, be many different systems of worlds–a strikingly modern view. Oresme also dared to compare the standard Earth-centered cosmos, as approved by the Church, with the heretical sun-centered scheme, concluding that either theory would serve to explain all the known facts, and that, therefore, it was impossible to choose between them. Even-handed and reasonable as this position may seem, it was a very hazardous one to take. Any questioning of the existing order was taken as an assault on the Church, which was quite prepared to defend its power and privileges with all the means at its disposal, including excommunication, torture, and, if necessary, the stake.

Yet, steadily and inexorably, the tide began to turn against centuries-old dogma. In the later Middle Ages, the rise of towns and trade saw the emergence of a new and vigorous element in the social equation. The burgeoning class of wealthy merchants began to flex its muscles and demand its rights. The expansion of commerce, the opening up of new trade routes, the rise of a money economy, the creation of new needs and the means of satisfying them, the development of arts and crafts, and the rise of a new national literature, all heralded the birth of a revolutionary force in society. As open-sea navigation developed, there was a need for new and better charts based on accurate astronomical observations.

No longer could the failings of the old Aristotelian scheme be ignored; no longer would they be tolerated. Neither the world nor the heavens behaved the way the Church insisted they did–and people were beginning to notice and speak up.

3

The Parabolic Man

On August 2, 1971, David Scott, commander of the Apollo 15 mission, carried out a little demonstration that would have amazed Aristotle and appalled any medieval philosopher or theologian. Standing on the moon–itself an inconceivable feat in the Aristotelian cosmos–he dropped two objects–a geological hammer and a falcon's feather (the lunar module was called *Falcon*)–at the same time from the same height. The feather didn't drift down, meanderingly, as it would have done on Earth. Instead, in the airless vacuum of space, it fell straight, without a flutter, keeping pace with the hammer and reaching the lunar surface at the same instant.

Even when we know what the outcome will be, even if we've seen the video of Scott's extraterrestrial physics primer a hundred times, it seems surprising and unnatural because it runs so contrary to the grain of everyday experience. Any child will tell you that, on Earth, feathers drop more slowly than stones, leaves tumble more lazily than branches. It's perfectly reasonable, as a working hypothesis, then, to assume that lighter objects in general always fall more slowly than heavier ones. When Aristotle made this claim, he was simply being guided by common sense and the evidence of his eyes. It's just a pity that no one bothered to check if he was right. The key experiments that finally put Aristotle's theory to the test–and to the sword–wouldn't come, astonishingly, for another 1,800 years. But when they did, they inspired a revolution because they opened the door to our modern understanding of gravity.

The Birth of a Myth

Galileo Galilei was born in Pisa in 1564, in the same year as Shakespeare, and died in 1642, the birth year of Isaac Newton, his scientific successor. His father was a music teacher and his family was of minor noble blood, though not wealthy. In 1581, Galileo began studying at the University of Pisa, where his father hoped he might pursue medicine. But mathematics and science, especially when coupled with a keen observation of nature, proved too powerful a lure for him. Legend has it that, while still a student, Galileo became intrigued by pendulums when he saw a suspended lamp swinging back and forth in the city's cathedral. Timing the swings with his own pulse, the story goes, he found that the period (the time in which the pendulum completes one trip back and forth) is independent of the arc of the swing. Grasping the importance of this to timekeeping, he later went on to develop a more accurate form of pendulum clock.

As were all his fellow students throughout Europe, Galileo was fed a stale diet of Aristotelian physics. But he also lived in an age of fast-receding horizons, of explorers heading into uncharted seas and returning with tales of wonder. There was a new spirit of questioning abroad, of wanting to see the truth of things for oneself. And so Galileo, who would have stood out as a genius in any century, became the first genius of experimental science. High among his priorities was to check, once and for all, Aristotle's claims about how objects move and fall.

Which brings us to a familiar tale. This short, stocky, red-haired man, now a very unconventional professor of mathematics at Pisa, who repeatedly runs afoul of the authorities for refusing to wear his academic gown while teaching, climbs the winding stairs of the Leaning Tower. From an upper balcony, he reaches out and lets fall two stones of different weights. A remarkable thing happens: to the gasps and amazement of the crowd gathered below, the stones hit the ground together.

Although doubtless in part apocryphal, this story does at least have some backing from Galileo's pupil and amanuensis Viviani, who reported that Galileo had done the experiment "in front of all the faculty and students assembled." Also, in his *Discourses on Two New Sciences* (1638), Galileo has one of his protagonists, Sagredo, say, "But I, Simplicio, who have made the test, can assure you that a cannonball weighing one or two hundred pounds, or even more, will not reach the ground by as much as a span ahead of a musket ball weighing only half a pound, provided both are dropped from a height of 200 cubits." Perhaps the tower experiment was really just a class demonstration, or perhaps it never happened at all and Galileo's description in *Two New Sciences* was no more than a mental exercise. Whatever the truth of the affair, it is unimportant scientifically because Galileo had already gathered all the data he needed about falling bodies from a different source: a series of beautifully crafted studies in which the pull of gravity is diluted and its effects made crystal clear.

Early Doubts

It's easy to get the impression that Galileo, a colorful, rambunctious, prickly character and one of the few scientists whom everyone can name, took on Aristotle and his institutionalized worldview virtually single-handed. Voices of discontent, however, had been raised very much earlier. Back in the fifth century A.D., the Greek Christian philosopher John Philoponus, also known as John the Grammarian (among other names), cast a critical eye over what Aristotle had to say in his *Physus.* Philoponus didn't buy Aristotle's explanation that an object in flight is nudged along by a vortex of air. Much more likely, he thought, a projectile keeps moving because of a "kinetic force" that is imparted to it by whatever sets it going (a hand or a bow, for example) and exhausts itself in the course of the movement. Similar ideas had been expressed even earlier, by Hipparchus in the second century B.C. and Synesius in the fourth century A.D. Although this theory of impetus, as it became known, was still flawed because it fudged the issue of how the kinetic force could work, it marked a crucial step away from Aristotelian dynamics toward a more modern theory of how things move–a theory based, as we'll see, on the concept of inertia.

The theory of impetus also gave Philoponus a new perspective on the role of the medium through which an object travels. Far from being the means by which the projectile is kept moving, the medium was seen in the impetus theory to be something that got in the way. This being the case, Philoponus concluded, there was nothing to prevent one from imagining motion taking place through a void and therefore no reason to disbelieve in empty space purely on the grounds that Aristotle defended. As for the natural motion of bodies falling through a medium, Aristotle's verdict that the speed is proportional to the weight of the moving bodies and indirectly proportional to the density of the medium was undermined by Philoponus through appeal to a mind's-eye conception of the same kind of experiment that Galileo would carry out centuries later.

A separate line of attack against Aristotle was taken up by Giovanni Benedetti, born in a very different era–just thirty-four years before Galileo. Starting his career as court mathematician to the duke of Parma, he moved to Turin at the invitation of the duke of Savoy to serve as philosopher-in-residence there. In his final and most important work, *Diversarum Speculationum* (Of Various Speculations, 1585), published around the time Galileo was finishing his college education, Benedetti drew attention to a thought experiment that effectively demolished Aristotle's theory of free fall. Imagine, said Benedetti, two equal weights connected by a gossamer thin, essentially weightless line. Thus joined, they must fall at the same rate as a single body having their combined weight. Now suppose the line is cut. Intuition demands that the two equal bodies, though disconnected, must continue to fall at the same speed they had before. And so Aristotle, who insisted that bodies with different weights came to earth at different rates, was undone by a simple flourish of logic. All that remained now was for someone to expose the untruth definitively in real life.

Inclined to Success

One of the difficulties of doing low-tech experiments with gravity is that objects drop vertically at a pretty brisk pace. There were no stop watches or high-speed cameras at the turn of the seventeenth century, so studying closely the behavior of falling bodies posed quite a problem; all the action was over before you could really tell what was going on. To this difficulty Galileo brought a brilliant and elegant solution. His realized he somehow had to slow down the rate at which an object fell so that he could accurately measure its speed. Yet he had to do this without altering the character of the motion. The trick he came up with was to replace the vertical drop of a body with a much more gradual roll down an inclined plane. This is how he describes his apparatus in *Two New Sciences*:

> A piece of wooden molding or scantling, about 12 cubits [some twenty-three feet] long, half a cubit [about one foot] wide and three finger-breadths [about two inches] thick, was taken; on its edge was cut a channel a little more than one finger in breadth; having made this groove very straight, smooth, and polished, and having lined it with parchment, also as smooth and polished as possible, we rolled along it a hard, smooth, and very round bronze ball.

The beauty of this setup, he pointed out, is that the speed gained in rolling down the ramp doesn't depend on its slope. A similar pattern would emerge, Galileo explained, if a ball rolled down a ramp that was smoothly connected to another steeper upward ramp; the ball would roll up the second ramp to a level essentially equal to the level it started at, even though the two ramps had different slopes. It would then continue to roll back and forth between the two ramps, before finally coming to rest because of friction.

Thinking about this motion, it's clear that if you ignore the gradual slowing down on successive passes, the ball must be going at the *same speed* coming off one ramp as it does coming off the other. Galileo then invites us to imagine the second ramp getting steeper and steeper until it becomes nearly vertical, at which point the ball is essentially in free fall. Thus, he concluded, for a ball rolling down a ramp, the speed at various heights is the same as the speed the ball would have reached, much sooner, by just falling vertically from its starting point to that height. This fact handed him the perfect solution: the physics of free fall could be modeled using a ball and a ramp. And by inclining the ramp at a gentle enough angle, the movement could be made sufficiently sedate to be measured. Effectively, Galileo had reduced the power of gravity so that he could watch objects descend in slow motion.

His only remaining problem was the lack of a precision timekeeper. And here again he showed his ingenuity by turning to an ancient device–the water clock, or *klepsydra*–by which he could actually weigh the moments of his experiments:

> [W]e employed a large vessel of water placed in an elevated position; to the bottom of this vessel was soldered a pipe of small diameter giving a thin jet of water, which we collected in a small glass during the time of each descent. . . . The water thus collected was weighed, after each descent, on a very accurate balance; the difference and ratios of those weights gave us the differences and ratios of the times.

Perhaps Galileo first acquired his skill as an empiricist while assisting his father, who carried out home experiments in the physics of sound. At any rate, Galileo was himself a very musical man with a natural sense of rhythm, and he put this to good use in evolving a second strategy for time measurement. In the path of the balls that he rolled down the plane he placed little bumps. Every time a ball went over a bump, it made a click. By arranging the bumps so that the clicks came in regular succession, he could be sure that the time between bumps was the same.

Galileo conducted his inclined plane experiments in hundreds of different ways to penetrate the mysteries of motion and fall. He varied the height of the plane and the angle it made with the horizontal. He varied the size and weight of the balls. In some trials the ball left the end of the ramp, which sat on a table, and descended directly in an arc to the floor. In other related experiments, a horizontal shelf was placed at the end of the ramp, and the ball would travel along this shelf before making its final plunge. Diligently, he measured the times and distances traveled, both horizontally and vertically, during each descent.

For Galileo, it wasn't enough to show beyond the shadow of a doubt, as he did, that objects of different weight fall at the same rate. He longed to know the precise law that governed falling bodies–the mathematics of unhindered descent. In this quest, his ear for tempo served him well. Listening for the clicks as each bronze ball rolled down the plane, he uncovered a wonderful secret of nature. In the first second of descent, the ball covered a distance of one unit. In the next second, it traveled three times this distance, and in the next second, five times the initial distance. This sequence spoke of uniform acceleration–the ball speeding up by equal amounts in equal times. Repeated tests, with the plane at different angles, always gave the same result: from one second to the next, as the ball rolled down the inclined plane, the ratios of the distances covered increased by odd numbers, that is, by intervals of 1, 3, 5, 7, 9, and so on.

After one second, the ball had covered a distance of one unit. In the next second, it covered three more units, so that at the end of the first two seconds it had traveled four units in total. In the third second, it covered five more units, for a total distance after three seconds of nine units. One, four, nine–the next in the sequence would be 16, then 25: these are, Galileo realized, the square numbers, 1^2, 2^2, 3^2, . . . Here was the intimate connection between time and distance for everything that descends under gravity: the distance an object falls varies as the square of the elapsed time. The precise formula took many years for Galileo to establish and verify and appeared in public for the first time in his *Two New Sciences*, more than three decades after the first inclined experiments were carried out. We can write it in the form $s = \frac{1}{2}at^2$, where s is distance, t is time, and a is a constant–the uniform acceleration due to gravity.

One of Galileo's greatest insights was to realize that a body's motion has two completely different and separate components. Movement in the vertical direction, if you ignore air resistance, is dictated by gravity and follows the time-squared rule just described. Horizontal motion, however, isn't affected by gravity at all. Aristotle claimed that it took some kind of push to keep going. But even in ancient times, this idea was refuted by Philoponus with his theory of impetus. Galileo's close contemporary Bennedetti then advanced the anti-Aristotelian argument further in a couple of ways. First, he claimed that the medium hinders, rather than aids, the motion, and second, he portrayed impetus as a quality transferred to a body that enables it to stay in motion. The longer the body is impressed with impetus, he said, the more it acquires. In Galileo's hands, this concept became still more refined. Impetus, Galileo explained, has similarities to both heat and sound. Once you transfer heat to a body, for example, it is hot and remains so until the heat has dissipated. If you strike a bell, it acquires a sonorous quality that continues even when the bell is no longer being disturbed. In the same way, motion imparted to an object remains until the medium resists and drains away the impetus.

Galileo explored impetus by thinking again in terms of a rolling ball and inclined planes. To begin with, a ball is rolling back and forth between two identical inclines. Suppose the ball and the surfaces are so smooth that there's no friction and therefore no loss of impetus over time. Whatever height the ball starts out from on the left incline, it climbs back up to on the right incline before reversing its motion. Now suppose the right incline isn't so steep. The ball again rises to the same height from which it was released, but now it has to roll a greater distance up the right incline before coming to a halt for an instant at the top of its journey. Therefore, it takes more time for the ball to roll to a stop on the right incline before it turns around. This time grows longer and longer as the slope of the right incline decreases. Finally, the right incline becomes flat. In this case, Galileo realized, the ball would take an infinitely long time to stop; in other words, it would carry on moving forever without any change in speed or course–unless something else, such as friction, affected it.

A Question of Curvature

These are some of the innovative ideas about vertical and horizontal motions that Galileo brought to bear on the old puzzle of determining the path of a projectile. Tartaglia had shown the way with his recognition, albeit hazy, that the path was curved everywhere, not mostly made of straight lines as Aristotle had taught. But Galileo was able to nail the actual curve–its precise shape and mathematical signature–by applying the theoretical and applied knowledge he'd gained from his inclined plane experiments. If a cannonball, obeying the time-squared law, falls vertically under gravity while simultaneously traveling horizontally at constant speed because of its impetus, its combined motions (ignoring air resistance and all other factors) will make it describe that most graceful of curves, the parabola.

Galileo may have known this fact from his experiments as early as 1608; he didn't publish his findings on projectiles until 1638, however, when they appeared in the second part of his monumental *Two New Sciences.* Six years earlier, his former student Bonaventura Cavalieri, a Jesuit, wrote a book called *Specchio Ustorio* (The Burning Mirror), in which he became the first to go to press with a mathematical proof of parabolic trajectories. Not surprisingly, having had his thunder and the results of three decades of careful research stolen–or heavily borrowed–in this way, Galileo wasn't amused. It says something of his magnanimity that he and Cavalieri were later reconciled after the younger man offered an apology.

The parabola is the curve that all projectiles trace out under the influence of gravity, *providing that the effects of the medium are negligible.* Galileo understood that air affects the path of an object, but he thought its influence could be safely ignored. As it happens, it can be ignored if the projectile is massive enough and moving fairly slowly. This explains why Galileo's calculated trajectories work so well for cannonballs and their ilk. As we saw in the previous chapter, ballisticians eventually went to all sorts of lengths to try to figure out the much more complicated paths of lighter, faster objects, such as bullets and artillery shells, for which drag is vastly more influential. But that's a different story.

Galileo had put mathematics into the heart of science, or natural philosophy as it was then still known. He had begun to uncover the laws by which things moved and fell. Through painstaking experiment and analysis, he had begun to dispel the mystery of gravity. Galileo's great genius lay in his ability to observe the world at hand, to understand the behavior of its parts, and, most tellingly, to describe what he found in terms of mathematical proportions. For these achievements, Albert Einstein dubbed him "the father of modern physics–indeed of modern science altogether."

And perhaps if Galileo had stopped there, with his studies of movement, gravity, and other terrestrial science, he would have brought less trouble on himself. But his vision wasn't confined to phenomena on or near the Earth. He dared to raise his eyes to the heavens and question what lay there–and that would put his very life at risk.

What Is and What Should Never Be

In 1600 the renegade Dominican friar Giordano Bruno was burned at the stake in Rome for offering views that differed rather too markedly from those of the Vatican. Among his heresies was to insist that the Earth didn't sit motionless at the center of the universe, but instead traveled around the sun and that–madness upon madness–"innumerable suns exist; innumerable Earths revolve about these suns. . . . Living beings inhabit these worlds." Few modern astrobiologists would disagree.

In the same year that Bruno's life was abruptly curtailed, that of Galileo's first child began. As Jacob Bronowski wrote in *The Ascent of Man*, Galileo "had rather more children than a bachelor should." His two daughters and one son were all born of his long illicit liaison with the beautiful Marina Gamba of Venice. The eldest offspring, christened Virginia, later became Sister Maria Celeste–the subject of Dava Sobel's best-selling book *Galileo's Daughter*–after entering the convent at San Matteo. Her sister, Livia, also took the veil and vows at San Matteo, while their brother, Vincenzio, the youngest progeny of Galileo and Marina, was eventually legitimized in a fiat by the grand duke of Tuscany and studied law at his father's old alma mater in Pisa.

Marina and the three children were living in Padua, and Galileo was teaching mathematics at the university there when the course of scientific history changed. In July 1609 Galileo learned of the invention of the telescope in Holland. Not knowing the exact details, he began working out the principles involved for himself, and by the beginning of 1610 he had built his first simple "optik tube." He set it up immediately in the garden behind his house, pointed it at objects in the night sky–the moon, the planets, the Milky Way–and was astounded by what he saw.

A bewildering number of stars leaped out at him, "more than ten times as many" as were visible to the naked eye. The Pleiades showed not merely the seven "sisters," long sung of by the poets, but thirty-six glittering members, while the faint haze of the Milky Way resolved into a breathtaking stellar host. The moon, far from appearing as the smooth orb the Scholastics supposed it to be, was "rough and uneven, covered everywhere, just like the Earth's surface, with huge prominences, deep valleys, and chasms." It was another world, in some ways resembling our own. With his primitive telescope trained on Jupiter, Galileo spied four attendant pinpoints of light. As he watched, from hour to hour and night to night, the lights changed position; they were, Galileo realized, satellites moving around their large parent, the whole comprising a planetary system in miniature.

Hastily, Galileo published his findings in a booklet called *Siderius Nuncius* (The Starry Messenger) in March 1610. "I render infinite thanks to God," he wrote after his wondrous nights at the eyepiece, "for being so kind as to make me alone the first observer of marvels kept hidden in obscurity for all previous centuries." Those marvels caused a sensation and were the talk of intellectuals and high society throughout Europe. They also transformed Galileo's life. He was soon appointed chief mathematician and philosopher to the grand duke of Florence and moved to that city to assume his position at the court of

Cosimo de' Medici. But even as his fame and prosperity grew, he attracted enmity and suspicion.

The planets went around the Earth, Aristotle and the Church maintained, so how could Jupiter have four private worlds of its own? The heavens were perfect and immutable, orthodoxy taught. Why then did the moon look rough and changeable, disturbingly familiar? Worse was to come.

Hardly had Galileo's first wave of cosmic revelations broken upon an astonished population than there was a new spate of discoveries. Galileo followed the progress of Venus, "the mother of loves," as he described it, week by week, and found something utterly astonishing. Venus cycled through phases from full to crescent, just as the moon did. And horror of horrors, the sun was blemished, marked by what Galileo called *macchie solari*–sunspots–that crawled continuously across its face. The perfect sun, the emblem of divinity, was speckled and despoiled.

Galileo was now sailing into dangerous waters, not just because of what he'd seen with his optically enhanced vision but because of how he interpreted those sights. For Galileo believed in the theory, espoused ages ago by Aristarchus but much more recently revived by the Polish cleric and amateur astronomer Nicolaus Copernicus, that the Earth trekked around the sun.

Revolutions

Copernicus, born in Torun, Poland, in 1473 and educated at Krakow University, had traveled to Italy in his youth, learned medicine and law there, and been infected with the new spirit of inquiry and free thinking that he encountered. Soon he found himself questioning the cosmological worldview of his Church superiors. There was something strange, he knew, about the motion of the planets: they didn't take part in the regular east-west procession of the other heavenly bodies. Mars, for example, after traveling east to west as expected, would pause in its motion for several nights and then mysteriously begin to travel backward from west to east, swimming against the heavenly tide. Several nights later, after this enigmatic excursion, it would resume its normal course from east to west. It was a puzzle that had troubled astronomers since antiquity: Why did the red planet trace out a loop in its journey across the sky?

Aristotle and, later, Ptolemy, had tried to account for the sometimes aberrant motions of the planets in their geocentric scheme by calling on an elaborate hierarchy of circular orbits. In the final, full-blown Ptolemaic model, each planet moves not only on a circular path around the Earth but also travels around a little "epicycle," designed to explain the occasional strange retrograde movements. Even this complicated arrangement doesn't work very well, but Copernicus realized that it works better if the object around which everything else revolves isn't the Earth but the sun. By the time he became canon of Frauenburg Cathedral in 1512, having succeeded his uncle in that post, Copernicus had abandoned the Ptolemaic system and began to formulate a new vision of the cosmos with the sun at its heart.

He first discussed his ideas in *Commentariolus*, a brief tract completed sometime before 1514 and circulated in manuscript form to interested scholars. Thereafter he fleshed out the details of his new system in *De Revolutionibus Orbium Coelestium* (On the Revolution of the Celestial Sphere). Although the manuscript was completed by 1530, Copernicus seems, probably for fear of papal retribution, to have been reluctant to publish it. In fact, it was a decade later before he was persuaded to do so, by the Austrian mystic and mathematician Rheticus, who was one of Copernicus's most outspoken advocates. The work finally appeared in 1543, just in time, according to popular legend, for it to be shown to Copernicus on his deathbed.

Although *De Revolutionibus* was banned by the Church, copies and word of the manuscript got around and began to shape the views of independent thinkers across Europe. A convert to Copernican theory, Galileo believed his startling telescopic observations offered it powerful support. How else to explain, for example, the shifting phases of Venus other than by assuming that it was in orbit close to the sun and was showing us various aspects of itself illuminated as it went around the central fire?

Galileo began to broadcast his ideas with great gusto–in bawdy humorous writings, loudly at dinner parties, and forcefully in staged debates. He became the leading spokesman for the new heliocentric alternative. That was a hazardous role to play. A committee of consultants declared to the Inquisition that the Copernican proposition that the sun is the center of the universe was a heresy. Although Galileo supported the Copernican system, he was warned by Cardinal Bellarmine, under orders from Pope Paul V, that he shouldn't discuss or defend that theory any further. Galileo knew the consequences if he failed to comply, so he fell silent on the subject.

For seven cautious years he channeled his efforts into less perilous pursuits, such as harnessing his Jovian satellites in the service of navigation to help sailors discover their longitude at sea. He studied poetry and wrote literary criticism. Modifying his telescope, he developed a compound microscope. "I have observed many tiny animals with great admiration," he reported, "among which the flea is quite horrible, the gnat and the moth very beautiful; and with great satisfaction I have seen how flies and other little animals can walk attached to mirrors, upside down."

But the siren of the new cosmos continued to call to Galileo, and in the summer of 1623, shortly after his sister's death, he found a reason to turn his attention back to the sun-centered universe. The old pope had died, and a friend of Galileo's, Cardinal Maffeo Barberini, had ascended the throne of Saint Peter to become Pope Urban VIII. Years earlier, Galileo had demonstrated his telescope to him, and the two had even taken the same side one night in a debate on the nature of buoyancy at the Florentine court. Urban, for his part, had shown his admiration for Galileo by writing a poem for him, mentioning the sights revealed by "Galileo's glass." He brought an intellectualism and an interest in scientific inquiry to the papacy not shared by his immediate predecessors, and he gave Galileo vague permission to write a book about how the planetary motions appeared to be. At the same time, it was made clear to Galileo that any treatment of Copernican theory was to be in terms of a mathematical proposition only. In the final analysis, there could be no doubting that the Earth stood at the center of the universe, just as the Church stood at the center of the world.

The temptation to speak his mind freely was too much for Galileo, however. In his *Dialogo Sopra i Due Massimi Sistemi del Mondo* (Dialogues on the Two Chief Systems of the World), published in 1632, he poked fun at the Church for its antiquated outlook. As in the case of his *Two New Sciences*, which came out six years later, the book was presented as a conversation between three men: the dull-witted Simplicio, who represents Church opinion and supports the Earth-centered hypothesis; Salviati, an intelligent man who stands for Galileo's views; and Sagredo, a wise and pragmatic man who is persuaded that Salviati's (Galileo's) views are correct. Simplicio's arguments in support of the Aristotelian worldview are invariably demolished by the other two, leaving the Copernican model effectively unchallenged.

Galileo had gone too far for his own good. At the age of seventy, and infirm, he was summoned to Rome to face the

Inquisition. Accused of heresy, he explained that his dialogue was merely trying to present all sides in the debate. But given his overt portrayal of Simplicio as an out-of-touch halfwit, this was scarcely a credible defense. His trial lasted two months, at the end of which the book was banned, and Galileo, under threat of torture, was forced to recant. "I abjure, curse and detest the aforesaid errors and heresies," he declared, "and I swear that in future I will never again say or assert, verbally or in writing, anything that might furnish occasion for a similar suspicion regarding me."

Thus Galileo avoided Bruno's fate, but at what cost? One of the greatest minds of the Renaissance was put under house arrest for the remainder of his life, Italian science was cowed and crippled, and the Church itself suffered grievously in the long run for its ruthless condemnation of a worldview that turned out to be correct.

Well, the Copernican theory was correct in one essential—that the sun, not the Earth, occupied the center of the planetary system. But in another sense, it was flawed. Its weakness, like that of the Aristotelian theory it opposed, lay in its insistence that the planets moved in *circles*, a combination of circular orbits and Ptolemaic epicycles. But just as a new curve had been introduced by Galileo to an understanding of projectiles, so another type of curve was needed to make sense of the planetary motions.

Galileo correctly envisioned the experimental, mathematical analysis of nature as the wave of the future: "There will be opened a gateway and a road to a large and excellent science," he predicted, "into which minds more piercing than mine shall penetrate to recesses still deeper." Among the first to bear out this prophecy was a man born within a year of Galileo's death—an unparalleled genius who codified the laws of motion and brought gravity to the universe.

4

The Day the Sky Fell

One had to be a Newton to notice that the moon is falling, when everyone sees that it doesn't fall.

—PAUL VALÉRY

No one knew quite what it was. Some said it was a displaced star. Others claimed it was a swath of flaming matter that might eventually fall into a dormant star and ignite it. The public was concerned that it might be an exhalation of dry air, portending a long spell of hot weather and possibly a spate of disease. It was a comet, an object that inspired both wonder and fear. Yet this particular comet, with its great silvery tail stretching across twelve degrees of the heavens, was special because, unlike any before it, it had been expected. Decades earlier, astronomers had foretold its appearance. And now here it was, brightening daily in the sky as 1758 drew to a close. The name it bore was Halley, but the cosmology its return testified to was that of the greatest scientific genius in history—the father of cosmic gravitation.

A Man Apart

Isaac Newton was born prematurely and sickly in the Lincolnshire hamlet of Woolsthorpe in England on Christmas Day 1642. His father, a farmer, had died a few months earlier, and within three years of Isaac's birth, his mother married again. Because his stepfather rejected the boy, he was left to be raised by his grandmother. Isaac's separation from his mother seems to have left him with a deep and permanent sense of insecurity, but one, curiously, that may have played a hand in some of his greatest discoveries.

Quiet, short-sighted, and reclusive as a boy, Newton showed from a young age those qualities of inquisitiveness and single-mindedness that would be the hallmark of his life's work. He is said to have made wooden models, including one of a mill powered by a mouse, and paper kites, which he flew at night with lanterns tied to their tails, terrifying the country folk nearby who thought they were comets. Fascinated by the motion of the sun, he marked on the ground with pegs the hours and half-hours cast by its shadow.

When Newton was fourteen, his mother, who had returned to Woolsthorpe after the death of her second husband, tried to make a farmer out of her recalcitrant son. After a few months, however, she gave up the unequal struggle and sent him back to the nearby Grantham School to prepare for a university education. At eighteen, he enrolled at Trinity College in Cambridge.

In appearance Newton was short, and the back of his head jutted out noticeably. His face was marked by a strong lower jaw, a broad forehead, and rather sharp features. His brown eyes had the absent look that comes from short-sightedness, and his hair went gray at thirty, remaining thick and silvery-white for the rest of his life. As a person he was remote and, although generally polite, rarely made close friends. He never married; indeed, it seems he died a virgin. His college notebooks reveal some of his habits. One of them contains, as well as entries for "a magnet, 16s.," "compasses, 2s.," "glass bubbles, 4s.," and "Philosophical Intelligences, 9s. 6d.," remarks such as "at the tavern several times, £1" and "lost at cards twice, 15s." He dressed carelessly and was languid in manner, often seeming completely detached from events around him. Once, riding home from Grantham, he got off his horse to lead it up a steep hill and then found, when he turned to remount, that although he had the bridle in his hand, his horse had slipped it and disappeared. He didn't exercise or go in for amusements but instead worked virtually nonstop, often spending eighteen or nineteen hours a day in writing. He took badly any kind of criticism. According to the astronomer John Flamsteed, who saw the worst side of him following a long, acrimonious feud, "He was of the most fearful, cautious, and suspicious temper that I ever knew." Newton's violent and vindictive attacks against friend and foe alike became increasingly common as he got older, and he lived his life on the verge of emotional collapse.

The early separation from his mother also appears to have left him perpetually longing for union–for the reconciliation that had eluded him in childhood and that was to escape him altogether when his mother died. Perhaps it's no coincidence that much of his intellectual effort went into a quest for oneness. In religion, he became a Unitarian; in optics he showed that white light is really the collective form of all the colors of the rainbow; and in cosmology, he effectively unified our understanding of the heavens and the Earth.

Like Einstein, Newton was a totally unremarkable student, uninterested in the fossilized physics and ancient math still being dished out to undergraduates at Cambridge. But his private notes make clear that he was deeply engrossed in far more advanced study than many of his peers, and he read the works of those spearheading the new scientific revolution–men such as René Descartes, Pierre Gassendi, and Thomas Hobbes.

During his early years at Cambridge, Newton pondered the question, How do the planets move in space? In order to begin to answer this, he assumed Copernicus was right about the sun being at the center of our planetary system. He also had at his disposal Galileo's discovery that a projectile fired from a cannon follows the curve of a parabola. And he knew about Kepler and his laws–vital clues, it turned out, in formulating the first complete theory of gravity.

The Stars and Their Courses

Johannes Kepler came from Weil der Stadt, in southwest Germany, and studied at the University of Tübingen. There he was taught by Michael Maestlin, who gave lectures to all comers on the Earth-centered Ptolemaic system, which he didn't really believe in, and tutorials to a few hand-picked students on the sun-centered scheme, which he did. In this clandestine way, Kepler came to know and appreciate the edgy new ideas of Copernicus. In 1597, at the age of twenty-four, he published *The Cosmographic Mystery*, in which, revealing his medieval mystical bent, he argued that the distances of the planets from the sun in the Copernican system were determined by the five regular, Platonic solids, if one supposed that a planet's orbit was circumscribed about one solid and inscribed in another. Except for Mercury, Kepler's construction presented–by sheer good luck, as it happens–surprisingly accurate results. The mathematical flair he showed in deriving his Platonic-based model of the planetary orbits drew the attention of a remarkable, larger-than-life character named Tycho Brahe, who, for the previous quarter century, had been amassing detailed observations of the night sky.

Born Tyge Ottesen, Tycho (his Latinized name) was the son of a distinguished Danish nobleman and later governor of Helsingborg Castle, opposite Elsinore, the town made famous in Shakespeare's *Hamlet.* Because he was an aristocrat, Tycho's choice of astronomy as a field of study, sparked when he saw a partial eclipse of the sun at age fourteen, was frowned upon; irrepressibly strong-willed, however, he brushed aside all family objections. After being sent off to the University of Leipzig in 1562 to study law, he carried on his celestial studies in secret; he then attended several more universities to focus on his scientific goal. The key to future astronomical progress, he realized, was continuous, accurate observation over long periods of time; larger, improved instruments were essential to carry through this program, and he, personally, was determined to make them and turn them on the heavens.

In 1566, at a wedding party, Tycho parted company with the bridge of his nose during a duel with his third cousin and fellow student, Manderup Parsberg, over who was the better mathematician. Subsequently, he wore a prosthetic nosepiece, purportedly made of pure gold, but which, upon examination by Czech scholars following Tycho's disinterment in 1901, proved to be a less-than-noble alloy of gold, silver, and copper.

During his studies in chemistry at Augustburg, Tycho persuaded his maternal uncle to install a laboratory near his castle at Herritzwad. Heading back from the lab for supper on the evening of November 11, 1572, Tycho caught sight of a brilliant "new star" in Cassiopeia, an object now known to have been a supernova. Surprised and excited, he first turned to his servants and some passing peasants to confirm what he'd seen. He then fetched a sextant of his own construction–a beautiful thing of walnut, with bronze hinges, its arms five and a half feet long, its gradations as accurate as sixteenth-century technology would allow. With this instrument he measured the distance of the stellar newcomer from the nine existing stars that make up Cassiopeia's familiar W-shape. Every night, when conditions allowed, Tycho recorded his observations of the new star as it gradually changed color and faded. They provided the subject matter for his first book, *De Nova Stella* (About the New Star), in 1573.

In the same year, his family relations became strained when he took a peasant girl as his common-law wife, possibly because he regarded his disfigurement as a bar to a socially correct marriage. He had intended to get away from his homeland and settle in Basel, Switzerland, but Frederik II, king of Denmark, bestowed upon him for life the beautiful island of Hveen, along with a hugely generous pension. Thus handsomely supported and accommodated, Brahe was able to set up the place of his dreams–a state-of-the-art observatory, built by a German architect under Tycho's supervision and known as Uraniborg (Castle of the Heavens).

Uraniborg emerged as a fusion of Tycho's meticulous precision and fantastic extravagance. Its centerpiece was an onion-shaped dome, flanked by cylindrical towers, each with a removable top–the dome and the towers–housing Tycho's instruments. In the basement was a printing press, an alchemist's furnace, and, shockingly, a private prison in which Tycho punished his servants and the peasants living on his lands when they broke one of his strict rules. The tyrannical Tycho, a law unto himself, also had his own paper mill, pharmacy, game preserves, and artificial fish ponds. Most significantly, he gathered or constructed some of the most fabulous astronomical instruments of his day–equipment that ranged from armillary spheres, after the style pioneered long ago by Ptolemy, to quadrants of bold and original design. Most of these devices were fashioned in his own workshop and on the grandest scale possible without sacrificing mechanical rigidity. Ample size and precise manufacture, Tycho had grasped, were the keys to achieving maximum accuracy in observation. His largest celestial globe, which graced the library, was five feet in diameter, made of brass, and engraved by Tycho with the fixed stars, after their correct positions had been determined by him and an assistant. His mightiest quadrant spanned fourteen feet and was fastened to a wall, the space inside its arc filled with a life-sized fresco of the autocratic astronomer amid his wonderful collection of paraphernalia.

For Tycho, painstaking observation was almost a form of worship. He rejected the Ptolemaic system because it didn't square with his data on planetary motions. But he didn't fully accept the Copernican model, either. Instead, he came up with his own hybrid scheme in which the Earth sits at the center of the cosmos, with the sun and the moon circling around it, and the other planets, in turn, revolving around the sun. One of his greatest contributions to astronomy was to smash the myth of Aristotle's crystalline spheres.

In November 1577, Tycho was fishing in one of the ponds in the gardens of Uraniborg when, just after sunset, he noticed a bright star in the west. As dusk fell, a splendid tail came into view, and Tycho realized that what he was looking at was no star but a brilliant comet. At once, he initiated a series of measurements with sextant and quadrant from different locations. These measurements revealed that the comet had no perceptible parallax–in other words, it showed no shift against the starry background when viewed by observers many miles apart; therefore, its distance had to be much greater than that of the moon, whose parallax was sizable. Tycho estimated that the comet's path lay outside that of Venus and in the direction opposite to that of the planets. If Aristotle's transparent spheres had been blocking the way, he realized, it would have been impossible for the comet to pass along its course. And so, he concluded, "There are no solid spheres in the heavens." Even Copernicus hadn't challenged that aspect of the classical cosmos; Tycho, far from the clutches of Rome, was the first to do so and thus to prove by observation that above our heads, as far as the distant stars, there is nothing but free and open space.

In 1588, Frederik II died and was succeeded by his son Christian IV. Tycho, already known as a despotic ruler on his private land, quarreled with the young king and didn't bother to answer his letters. He ignored provincial as well as high court rulings by holding a tenant and his family in chains. Christian IV didn't bring any direct action against him, but he did cut the astronomer's princely income to a more reasonable level. It was too much of an insult for Tycho, who, in any case, had grown restless and bored in his island paradise. In 1597 he left Hveen with more luggage than a rock band: his printing press, his library, his furniture, and all his instruments, except the four largest ones, which followed later, in his train.

From the outset Tycho had ensured that all his technical gear, despite its size, was made so that it could be dismantled and taken from one place to another. For two years he wandered around Europe before arriving in June 1599 in Prague. There he found a new patron in Rudolf II, King of Hungary and Bohemia as well as Holy Roman Emperor, who appointed him imperial mathematicus of the empire, the most prestigious job in mathematics. Again Tycho found himself the beneficiary of a munificent annual salary–some three thousand florins–and the residence of his choice, Benatky Castle, several hours by carriage northeast of the capital. It was here, in 1600, that

Tycho, having been impressed with the work of Johannes Kepler, invited the young German to be his new assistant.

They must have seemed an unlikely pair, the legendary, bluff astronomer, burly of figure, his plump face sporting a six-inch-long drooping mustache, opulently bedecked with lace and jewels, and the poor innkeeper's son and former theology student, with his threadbare clothes, thin frame, and tense, timid manner. Despite some disagreements about the truth of the Copernican theory, for example, they forged a successful, albeit brief, partnership, and Kepler set to work calculating new orbits for the planets based on Tycho's voluminous records.

The following year Tycho died, not of a burst bladder following a gastronomic orgy, as some accounts suggest, but from high levels of mercury (which he may have taken as medication after falling ill from his infamously large meals) in his blood. Kepler succeeded him as imperial mathematicus and launched into a career that would change forever our view of the solar system and provide Newton with the platform he needed for his spectacular breakthroughs in gravity.

The Shape That Shouldn't Be

Kepler, like many of his contemporaries, had a foot in both worlds–the ancient and the new. He subscribed to the heliocentric theory of Copernicus yet believed with all his heart that God had created the cosmos according to perfect geometric figures and mathematical proportions. That's why he felt moved to place the planetary orbits on the boundaries of Platonic solids. It was inconceivable to him that the orbits themselves could be anything other than circles, or circles within circles. When he began sifting through Tycho's data, he felt sure they would vindicate the age-old conception of the celestial spheres, with or without Aristotle's adamantine crystal. Yet the data stubbornly refused to conform to this picture. Kepler imagined himself on Mars and tried to reconstruct the Earth's motion from that vantage point. Nine hundred pages of calculations later, Kepler was driven to a fantastic, almost unthinkable conclusion. The orbits of the planets were not circular at all; they were *elliptical*, with the sun at one of the focal points. (An ellipse has two focal points that lie on its longest axis, such that the total distance from one focal point to any point on the circumference of the ellipse and then to the other focal point is a constant.) From Tycho's observations, Kepler was also able to deduce that the line joining a planet to the sun sweeps out equal areas in equal amounts of time, no matter where in its orbit the planet may be. Both these discoveries were published in his *Astronomia Nova* (New Astronomy) in 1609 and subsequently became known as Kepler's first and second laws of planetary motion.

A third law, stating that the square of the period of revolution of a planet varies as the cube of its average distance from the sun, appeared a decade later in *Harmonices Mundi* (Harmonies of the World). Ironically, Kepler stumbled upon it while trying to show that each planet has a distinctive range of voice–bass, tenor, contralto, and so on–and that the music it gives off is determined by its distance from the common center of revolution, the sun, just as the length of a string on an instrument determines its sound. That notion of "the music of the spheres," which we still occasionally invoke when pondering the stark beauty of outer space, goes all the way back to the Pythagorean preoccupation with sound and shows itself also in the belief of Philolaus, a leading Pythagorean and the teacher of Archytas, that the spheres of the various planets made celestial music as they turned. Strange, that in harking back to such an old and outmoded theme, Kepler should uncover a law so breathtakingly modern.

His propositions struck a heavy blow against the orthodox positions of the Church. Nor was Kepler himself at all pleased with them. How could something so hopelessly imperfect as an ellipse be compatible with the idea of divine harmony of the universe? He consoled himself by regarding his discovery as "one more cart-load of dung as the price for ridding the system of a vaster amount of dung."

Kepler's insights–his three laws of planetary motion–exploded upon the western world like a thunderclap. When his laws were superimposed upon the orbits of the planets, all the complex Ptolemaic retrogressions and filigreed epicycles disappeared, as did the similarly burdensome complexities of the circle-based Copernican scheme. All that remained were clean, smooth elliptical paths around the sun. Kepler, reluctantly, almost regretfully, had unlocked the stupendous secret of the heavens.

Kepler had found the rules by which the planets moved, but he didn't know *how* they moved. He supposed that the sun swept them along almost as though with an invisible broom. The heavens remained a place apart, less perfect, perhaps, than had been supposed but still governed by rules that were not of this world. That status, however, was about to change.

Fruitful Times

The year was 1665, and Isaac Newton had just received his bachelor's degree at Cambridge without honors or distinction. The Great Plague–the Black Death–was sweeping across Europe and had reached the shores of Britain; in that terrible summer of 1665, more than one in seven of the English population died from it. Cambridge University closed its doors for almost two years, and in that time Newton, who had returned home to Woolsthorpe Manor, began to rewrite the science of physics. These plague years were his wonder years, for, as he recalled, "I was in my prime of age for invention, and minded mathematics and philosophy more than at any time since." In mathematics he conceived an early form of calculus, or what he called his method of fluxions, laid the foundations for his great opus on light and color, and began to penetrate the problem of planetary motion.

The tale of Newton's apple is as iconic in our collective imagination as that of Galileo's experiment from the Leaning Tower–and just as shrouded in myth. One evening in 1666, the story goes, Newton was sitting under an apple tree at Woolsthorpe, deep in reverie. As he watched the moon float above the horizon, one of the fruits fell to the ground. In that instant, he was inspired to wonder about the universality of gravity. If apples tumble to the Earth, why doesn't the moon fall also? Science is made to seem neat and tidy by belief in such epiphanies, and there's added attraction in the drama of supposed sudden breakthroughs. But the truth is usually messier and a little less romantic, and changes in worldview are more protracted.

Newton would later write that in 1666, "I began to think of gravity extending to the orb of the moon." His early manuscripts suggest, however, that in the plague years, any notions he had about gravity being a cosmic force were hazy and embryonic. The fact is that Newton's groundbreaking work on dynamics and gravity spanned two decades, a testimony not only to the complexity of his achievement but also to the drawn-out character of scientific discovery.

The years 1665 and 1666 mark the first phase of Newton's investigations into how bodies move and the nature of gravity. Out of them emerged early versions of his three laws of motion (not to be confused with Kepler's three laws): (1) An object will continue in a state of rest or uniform motion unless acted upon by a force; (2) force is proportional to the rate of change of momentum; and (3) to every action there is an equal and opposite reaction. Galileo had already stated the first law in a different guise (now usually referred to as the law of inertia). He also discovered the formula describing the centrifugal force on a body moving uniformly in a circular path: the centrifugal force is proportional to the square of the body's velocity and inversely proportional to the radius of its orbit.

Regarding gravity, Newton's big idea of 1666 was to imagine that the Earth's gravity influenced the moon, counterbalancing its centrifugal force. That was a remarkable insight. The moon moves in space without ever falling to the Earth; an apple falls, accelerates, and thuds to the ground. Only a genius or a fool would suspect that in both cases a common phenomenon might be at work. In those crucial years at Woolsthorpe, however, Newton began to suspect that there was a connection. To the question "Why doesn't the moon fall to the Earth?" he gave a surprising answer: the moon *does* fall. It falls continuously toward the Earth *but keeps missing it.* For this to be the case, Newton asked, what force of attraction between the Earth and the moon must there be?

He focused on the moon's centrifugal tendency–what he called its centrifugal "endeavor"–to recede; in other words, he considered the fact that if there were nothing to restrain the moon, it would fly away from the Earth. Evidently some counterpull, equal and opposite, to the centrifugal endeavor was present, bending the moon to its path. Although Newton wasn't yet thinking in terms of forces acting at a distance, he wondered if gravity–whatever it was–might be the common factor that caused a falling apple and a revolving moon to move as they do. He knew about Kepler's third law of planetary motion, which states that the square of a planet's orbital period is proportional to the cube of its mean distance from the sun, but he did no work at this stage that had any bearing on elliptical orbits–the subject of the first and second laws. Taking the moon's orbit to be circular, and combining Kepler's third law with his own newly conceived formula for centrifugal force, he made a remarkable discovery: the constraining force acting on the moon, his analysis showed, varies as one over the square of the orbital radius.

The next step was to test the inverse square relation he had just derived against empirical data. To do this, Newton, in effect, compared the restraint on the moon's "endeavour" to recede with the observed rate at which falling objects accelerated on Earth. If gravity underpinned both, then the moon and an apple, if dropped from the same height, should fall at the same rate. The problem was to obtain accurate data. Galileo had estimated that the moon is 60 Earth radii from the Earth, so the restraint on the moon should have been 1/3,600 ($^1/_{60^2}$) of the gravitational acceleration on Earth. It was known that a body on (or very close to) the Earth's surface falls in such a way that its speed increases by 32 feet per second every second: in the first second it falls 16 feet, during the next second it falls 48 feet, and so on.

To figure out the rate of the moon's fall to Earth, Newton needed the exact radius of the Earth, which, in turn, could be calculated from the size of a degree of latitude. Because he was away from Cambridge and his books, Newton used the most current value he could find: 60 miles to one degree of latitude. But this, it turns out, was about one-sixth too little. As a result, Newton calculated the effect on the moon to be about 1/4,000 of that on Earth. This was close to the expected value of 1/3,600, and, as Newton later described it, the moon test answered "pretty nearly." But it wasn't close enough. Although the theory would eventually prove to be correct, Newton had been let down by inaccurate data, and he abandoned the problem for more than a decade.

Difficult Relationships

In April 1667 Newton returned to Cambridge and, against stiff odds, was elected a minor fellow at Trinity. In the next year, upon receiving his master of arts degree, he became a senior fellow, and in 1669, before he'd reached his twenty-seventh birthday, he succeeded his old tutor, Isaac Barrow, in the coveted position of Lucasian Professor of Mathematics.

There's a temptation to think of Newton as the first of the modern scientists. The phrase *Newtonian mechanics* is often used to describe the science that, once and for all, swept away the old Aristotelian dogma. Yet Newton himself was really the last of the old guard. He still practiced astrology and believed in the power of alchemy. In 1678 he suffered a serious emotional breakdown, and in the following year his mother died. His response was to cut off contact with others and immerse himself in alchemical research. Hiding behind the pseudonym Jeova Sanctus Unus (God's Holy One), he wrote in his notebooks of ethereal spirits and a secret fire pervading matter. In quicksilver–mercury–he saw "the masculine and feminine semens . . . fixed and volatile, the Serpents around the Caduceus, the Dragons of Flammel."

What part such esoteric musings played in shaping his scientific worldview isn't easy to fathom. But there's no doubt that Newton's alchemical studies opened theoretical avenues not found in the mechanical philosophy–the worldview that sustained his early work. While the mechanical philosophy, which he shared with many great thinkers going back to the Greeks, reduced all phenomena to the impact of matter in motion, the alchemical tradition upheld the possibility of attraction and repulsion at the particulate level. It may be that Newton's later insights in celestial mechanics are traceable in part to his alchemical interests.

In any event, his excursions into the realm of the Philosopher's Stone were interrupted in November 1679 by a letter he received from Robert Hooke. An extraordinarily fertile inventor and source of new ideas–and perhaps the greatest English experimentalist before Michael Faraday in the nineteenth century–Hooke lacked the technical gifts of Newton and some of his other contemporaries; otherwise he might have crafted more comprehensive theories. This weakness often led him into bitter disputes because he would claim that his original concepts had been stolen and elaborated on by his rivals. He had already come into conflict with Newton in 1672, for example, following the publication of Newton's theories of color and light. Hooke insisted that what was correct in Newton's thesis was plundered from his own ideas about light, arrived at seven years earlier, and what was original was wrong. Now Hooke wrote to Newton asking his opinion "of compounding the celestiall motions of the planetts of a direct motion by the tangent and an attractive motion towards the centrall body. . . . [M]y supposition is that the Attraction always is in a duplicate proportion to the Distance from the Center Reciprocall."

The second of these statements refers to an inverse square relationship–exactly the kind of law that Newton had derived in the plague years and then set aside because the moon test data he used didn't quite bear it out. The first statement of Hooke's, however, hints at something new, and in ensuing correspondence, Hooke specified more clearly what he had in mind. There's a central attracting *force*, he believed, that falls off with the square of the distance. This was the vital conceptual leap Newton needed to be able to crack the problem of planetary orbits. When he had analyzed the orbital motion earlier, his attention had been on centrifugal tendencies. But now he realized that the key lay in central attraction–the pull of a central force that continuously diverted the orbiting body from what would otherwise have been a straight-line path. The written exchanges with Hooke in 1679 mark the start of a new phase in Newton's gravitational studies. Having seen the way ahead, Newton abruptly broke off the correspondence, and, sometime in early 1680, he began quietly and alone to push toward the brink of an all-embracing gravitational synthesis. First he proved that Kepler's second law followed directly from the assumption that what holds a body in orbit is a central gravitating mass. Then he showed that if the orbital curve is an ellipse under the action of central forces, the radial dependence of the force is inverse square with the distance from the center.

In June 1682 at a meeting of the Royal Society, Newton heard about the work of Jean Picard, who was mapping France with sophisticated instruments and had found the length of a degree to be 69.1 miles. Repeating his former moon test calculations using Picard's value, Newton found that the rate of the moon's fall to Earth exactly corresponded with that predicted by the inverse square law. All the pieces of the puzzle of planetary motion were beginning to fall into place.

No one as yet had an inkling of the tremendous progress Newton had made in bringing gravity to heel. He rarely published anything promptly, and he wasn't about to announce a discovery that might leave him open to criticism. In the coffee houses of London, Robert Hooke and two other leading intellects of the day, the astronomer Edmund Halley and the architect Christopher Wren, met and wrestled with the very problem that Newton had already solved. Finally, in August 1684, Halley paid a visit to the great man in Cambridge, hoping for an answer to his riddle: what type of curve does a planet follow in its orbit around the sun, assuming an inverse square law of attraction? To Halley's joy and amazement, Newton replied, without hesitation, *an ellipse.* When asked how he knew, Newton said that he'd already calculated it.

In private and without telling anyone, Newton had solved one of the great conundrums of the age. He alone possessed the mathematical tools to have done so because the solution rested on his own brilliantly conceived method of fluxions. The key was to be able to study an orbit not as something frozen and predetermined but as an entity that varied continuously, like a flowing river. Newton considered the motion along an orbit from one point to another during an infinitesimally small interval of time and worked out the deflection from the tangent (a line just touching the curve) during that interval, assuming the deflection to vary as the inverse square of the distance from the center of motion. In this way, he proved, mathematically, beyond any possibility of doubt, that the curve a planet follows under the sun's attraction, or that the moon follows around the Earth, is an ellipse.

Unfortunately, and again this was characteristic of the man, he had misplaced the calculation. Halley, however, was not to be put off so easily. Dynamic, ebullient, charming, and diplomatic, he was the very antithesis of Newton, who at forty-one was fourteen years his senior. Before the young man left, Newton had promised to track down the errant papers and send them to Halley in London. His search proved unsuccessful, and he had to do the calculations again. When Halley eventually received them a few months later, he immediately grasped their importance and again traveled to Cambridge. This time he was given what Halley called "a curious treatise," *De motu corporum in gyrum* (On the motion of bodies in an orbit). Halley was so excited by what he read that he urged Newton to set down all his main mathematical arguments, proofs and conclusions to do with motion and gravity, together with their implications for astronomy, in the form of a book. Initially, Newton refused, but Halley persisted, bringing his outstanding powers of persuasion to bear.

Just as Rheticus had urged Copernicus to publish *De Revolutionibus*, so Halley pressed Newton, using just the right blend of tact, urgency, and discreet flattery, and promising to take care of all the practical arrangements. In the end, Newton agreed, and Halley, true to his word, personally financed Newton's magnum opus and guided it safely through a minefield of potential hazards. Not the least of these took the form of Robert Hooke's claim, with some justification, that his letters of 1679 to 1680 had earned him the right to a bit of recognition in Newton's discoveries. Newton was so furious with Hooke for trying to muscle in on what he considered his personal accomplishments that he threatened to suppress the final part of the book altogether, denouncing science as "an impertinently litigious lady." Although Newton finally calmed down under Halley's soothing influence, he wouldn't acknowledge Hooke's contribution; on the contrary, he systematically deleted every possible mention of Hooke's name from his work.

Principia

In July 1687, after nearly two years of intense labor, *Philosophiae Naturalis Principia Mathematica* (Mathematical Principles of Natural Philosophy)–or *Principia*, as it's known–was published and dedicated to the Royal Society. Widely considered the most significant printed document in the history of science, Newton's masterpiece is divided into three books.

The first book opens with a series of definitions and three axioms that later became known as Newton's laws of motion: the law of inertia, the law of proportionality of force and velocity, and the law of equality of action and counteraction. On these foundations, Newton builds the basic principles of a new vision of motion and mechanics–the core of what would be called classical physics.

For the first time, the concept of *inertia* (from the Latin *in* + *ars*, hence *iners*, meaning "unskilled" or "artless") acquires its modern meaning. Kepler had applied the word in a physical sense but used it only for bodies at rest. Galileo discovered what became known as the law of inertia, but he didn't refer to it by that name. In *Principia*, however, Newton wrote, "A body, from the inert state of matter, is not without difficulty put out of its state of rest or motion. Upon which account, this *vis insita* may, by a most significant name, be called *vis inertia*, or force of inactivity." Also in Book I he talks about orbital motion and shows that gravitation is the only force needed to account for the elliptical orbits of the planets and their satellites.

In Book II he tackles the motion of bodies through resisting mediums, as well as the motion of fluids themselves. Since Book II wasn't part of Newton's initial outline, it has traditionally seemed somewhat out of place. But, in fact, it segues very neatly into the crucial final book.

In 1644 the French philosopher and mathematician René Descartes had put forward some ideas of his own about the mechanics of the solar system in his so-called theory of vortices. Because Descartes, like Aristotle, believed that forces can act only when they're in direct contact with objects, he dismissed the notion of a vacuum in space. Instead, he assumed that the universe is filled with matter, which, due to some initial motion, has settled down into a system of eddies that carry the sun, the stars, the planets, and comets in their paths. Near the end of Book II Newton demolishes Descartes' theory by showing that there's no way the vortices could sustain themselves and that, in any case, the theory isn't consistent with Kepler's three planetary rules. The purpose of Book II then becomes clear. After disposing of Descartes, Newton concludes, "How these motions are performed in free space without vortices, may be understood by the first book; and I shall now more fully treat of it in the following book."

Immodestly but accurately subtitled *The System of the World*, Book III does nothing less than put gravitation on a cosmic stage. By combining his mathematics of motion with the idea of action at a distance, Newton finally demonstrates "that there is a power of gravity tending to all bodies, proportional to the several quantities of matter which they contain." Specifically, he says, "all matter attracts all other matter with a force proportional to the product of their masses and inversely proportional to the square of the distance between them."

This statement, which in time became known as Newton's law of universal gravitation, reduces in mathematical shorthand to $F = GMm/r^2$, where F is the gravitational force between two objects with masses M and m, and r is the distance between them. G is a constant (called the *universal gravitational constant*) whose value depends on the units used for mass and distance.

In formulating his law, Newton, without referring to it as such, introduced the idea of a force field–a region of space in which a force operates without the need to be in contact with objects that are affected by the field. The implications of such action-at-a-distance were far-reaching, literally and figuratively, and would have seemed quite alien to philosophers of an earlier generation. Newton's inverse square equation told how bodies separated by nothing but empty space could influence each other. Two billiard balls, for instance, placed a million miles apart in an otherwise deserted corner of the universe, will, according to Newton's formula, gradually pull each other together and eventually, after a fantastically long period of time, collide. Gravity is the force that mediates this strange interaction, causing inanimate objects to be drawn to one another despite the presence of nothing between them.

To demonstrate the power of his theory, Newton used it to explain the orbits of the planets and their satellites, as well as the motion of our own moon, including all the subtle variations in it caused by the tugging, this way and that, of the sun and the Earth. He accounted for the unusual paths of comets. And, most impressively, he offered an explanation for the ebb and flow of the tides. Until *Principia*, the tides were presumed to be caused by the moon alone, but no one quite knew how; Descartes, for instance, thought the moon exerted pressure downward on the seas. Newton pointed out that both the moon and the sun caused the seas to be pulled upward.

In *Principia*, for the first time, we see not one rulebook for the Earth and another for the heavens but a universe united by a single set of laws. The same force of gravity that brought apples down from trees and guided the arc of a flying cannonball also caused the planets to wheel in their endless treks around the sun. The cosmos would never seem the same again.

The War of the Worlds

Following the publication of *Principia*, Newton drifted away from science and became more involved in public affairs. In 1689 he was elected to represent Cambridge in Parliament, and during his stay in London, he became acquainted with John Locke, the philosopher, and formed a particularly close friendship with Nicolas Fatio de Duillier, a brilliant young Swiss mathematician. In 1693, however, Newton suffered another nervous breakdown, possibly due to overwork, Fatio's move back to mainland Europe, or chronic mercury poisoning–the result of nearly three decades of alchemical research. Whatever the cause, shortly after his recovery, Newton sought a new position in London, and in 1696, with the help of Charles Montague, a fellow of Trinity and later the earl of Halifax, he was appointed warden and then master of the Mint. Although his most creative years were behind him, Newton continued to have a huge influence on the development of science. In 1703 he was elected president of the Royal Society and thereafter annually reelected until his death in 1727. By that time, Newton's law of universal gravitation was being taught in all English universities.

On the Continent, however, it was a different story. The French continued to favor Descartes' vortex theory, in which forces work through contact. The idea of action at a distance–of a pull of gravity across empty space–smacked to them of mystification, on a par with Aristotle's belief that objects seek their natural place. As the *Journal des Savants* saw it, the mechanics of *Principia* "do not fulfill the necessary requirements of rendering the universe intelligible."

Acceptance of Newton's theory in France and elsewhere in Europe was further held back by Bernard de Fontanelle's whimsical *Entretiens sur la Pluralité des Mondes* (Conversations on the Plurality of Worlds), first published in 1686. Although not a scientist himself, Fontanelle had an elegant and compelling literary style that helped him become, effectively, the first science popularizer. Voltaire considered him "the most universal genius that the age of Louis XIV has produced." He was also a big fan of Descartes' vortices and used them to argue his case that life was common throughout the solar system and that inhabited planets went around other stars. Like many who followed him, he found that extravagant claims about extraterrestrial life sold well; *Conversations* was eventually translated into every major European language and influenced the thinking of generations to come.

It was Voltaire himself who began to swing Gallic public

opinion the other way. During his first youthful visit to England, he had become a staunch supporter of Newtonianism, and, upon returning home, he collaborated with his mistress, Emilie du Châtelet, in writing a lucid, popular account of the scientist and his work. Published in 1737 as *Éléments de la Philosophie de Newton* (Elements of Newton's Philosophy), it explained the new vision of gravity from across the Channel in a way that anyone with a lively, inquiring mind could grasp. Emilie then went on to prepare a French translation of *Principia*, which Voltaire had published after her death.

By this time, popular expositions of Newtonian science were coming thick and fast, feeding a growing market among the upper and middle classes for fashionable intellectual topics of conversation. Francesco Algarotti's *Newtonianism for Ladies*, chock full of amusing digressions, used the format of a genteel dialogue between a chevalier and a marchioness, as Fontanelle's *Conversations* had also done, to appeal especially to a female audience.

In Germany, for a while, the most important home-grown rival to Newton's theory of gravity was supplied by Gottfried Leibniz. These two men had clashed during life on the issue of whose version of the branch of mathematics now called calculus–Newton's method of fluxions or Leibniz's "differentials"–had priority. (Modern mathematicians tend to regard this contest as ending more or less in a tie.) As for their philosophical and cosmological views, Newton and Leibniz were poles apart, the German having decided that the universe is made up of countless tiny conscious centers of spiritual force or energy known as monads. These little thinking entities are constantly pushing one another around, but always in harmony, so that eventually they give rise to the most perfect of logically consistent structures.

"All is for the best in the best of all possible worlds," wrote

Leibniz, a view that Voltaire lambasted in *Candide*. Though he believed in the basic, underlying unity of the human spirit and had tried to reunite the Protestant and Catholic Churches, Leibniz wasn't able to work out a compromise between his own conscious monads and Newton's lifeless matter spinning in space. By the middle of the eighteenth century, most educated people across Europe had come around to accepting the cosmos in Newtonian terms–the result of the mutual attraction of bodies and the matter they contain.

The final triumph of Newton's theory of gravity came in the most dramatic and fitting style imaginable. When a brilliant comet lit up the sky in the autumn of 1682, Halley plotted its movements and noticed strong similarities between its course and those of two other comets seen in 1537 and 1607. He concluded that all three sightings were, in fact, the same object. The French mathematician Alexis Clairaut then took up the challenge of predicting when the comet would next come back to the vicinity of the sun.

Using Newton's principles, Clairaut worked out an orbit for the body that took into account the gravitational nudges that the comet received from the most massive planets, Jupiter and Saturn, as it passed them by. With these perturbations factored in, Clairaut forecast that the comet would next reach perihelion–the point in its orbit nearest the sun–on April 15, 1759, with an uncertainty of one month on either side of this date. Sure enough, astronomers first spotted the return of the long-distance traveler on Christmas Day 1758, and perihelion was attained on March 13, 1759, just at the margin of Clairaut's prediction. Newton's theory had been vindicated, and although no one yet understood what comets really were, their movements, at least, could now be foretold using the same mathematical rules as those that governed the orbiting of planets and the falling of fruit.

Close Encounters

On March 13, 1986, Halley's comet had a visitor–a small cylindrical thing, black, white, and silver, approaching fast. Eight months earlier, this strange object had been shot from the surface of its creators' world. Its name was Giotto, chosen because the Italian artist and sculptor Giotto di Bondone had been the first to accurately depict the famous comet in his fresco *Adorazione dei Magi* (Adoration of the Magi), painted sometime between 1304 and 1306. Almost certainly, Giotto had seen the comet firsthand during its appearance in the skies over Europe in October 1301. In 1986, the spacecraft Giotto closed in on its target–the heart of Halley, the comet's small solid nucleus–at a relative speed of 150,000 miles per hour. Closing further, all its instruments alive, it sent back images and other data. Giotto flew to within two hundred miles of the nucleus, mercilessly blasted by debris from the comet. Even its tough Kevlar shield could not protect it from every strike. A couple of seconds before its closest approach, Giotto's communications antenna was knocked out of alignment with Earth and its camera smashed, but not before the probe snapped a sensational view of the ice-rock core spraying dust and vapor into space.

Newton had shown, through his universal theory of gravity, that the heavens and Earth were not governed by different laws. The rules by which the planets moved and objects fell were one and the same. Through his theory, we could understand even the eccentricities of comets–and their orbits. And in time, just as Newtonian gravity had brought the skies down to Earth, so its principles would allow us to send our machines and ourselves to explore the boundless frontiers of space.

5

Escape from Earth

We can lick gravity, but sometimes the paperwork is overwhelming.

—WERNHER VON BRAUN

We live at the bottom of a well from which escape seems impossible. Jump up, and gravity quickly brings you back to terra firma. Throw a stone, and, however much effort you put into it, the stone traces out an elegant, near-parabolic arc before thudding to the ground a few seconds later. The deep well of gravity in which we spend our lives ensures that whatever goes up must eventually come down. Or perhaps not.

On October 5, 1957, people everywhere awoke to the stunning news that not all human-made things were now in their home world. The previous day, the Soviet Union had fired its Sputnik 1 probe into Earth orbit atop a rocket that could equally well deliver a nuclear warhead. Gravity had been defeated–at least temporarily. The two-foot-diameter, 185-pound sphere circled our planet for six months before tumbling back through the atmosphere. It was the first small step for mankind in freeing itself of its gravitational chains–and the realization of an ancient dream.

The Man in the Moon

One of the first visionaries with the nerve to imagine that space travel might be possible lived in the second century A.D. and came from Samosata, on the banks of the Euphrates in what is now northern Syria. He was Lucian–a wit, a raconteur, and one of the outstanding satirists of the ancient world. Lucian's parents had hoped he might become a sculptor, but instead, he made a fortune by doing his own thing: traveling around Asia Minor, Greece, Italy, and other lands, giving entertaining speeches, before settling down in Athens to study philosophy. He lived at a time when faith in the old gods had all but evaporated, Greek culture and thought were in decay, and the great classical literature of Greece had given way to shallow novels of adventure and romance.

All this was grist for Lucian's satirical mill, and in his two extraterrestrial yarns–the earliest forerunners of science fiction–he parodied the kind of feeble fantasy that had become popular. Concluding the preface to his mischievously titled *True History*, he wrote, "I give my readers warning, therefore, not to believe me." And with that, he launched into a tale of a group of adventurers who, while sailing through the Pillars of Hercules (the Strait of Gibraltar), are lifted up by a giant waterspout and deposited on the moon.

In his second space outing, *Icaro-Menippus*, Lucian was again Luna-bound, but this time in the footsteps, or rather the wing-flaps, of his hero, who has improved on the ill-fated scheme of

Icarus and his thermally suspect wax. To his incredulous friend Menippus, the intrepid voyager explains, "I took, you know, a very large eagle, and a vulture also, one of the strongest I could get, and cut off their wings." Like many who followed him, Lucian didn't distinguish between *aero*nautics and *astro*nautics, assuming that normal air-assisted flight and breathing were perfectly possible on jaunts between worlds.

All kinds of ingenious and eccentric schemes were hatched over the centuries by writers eager to whisk their travelers away from this planet's embrace. In 1532 the Italian poet Ludovico Ariosto penned an epic poem, "Orlando Furioso," in which one of the characters flies to the moon in the same divine chariot that carried the prophet Elijah in a whirlwind to heaven. Johannes Kepler, whose laws of planetary motion would eventually open the door to real space exploration, wrote a fantastic story called *Somnium* (The Dream), published posthumously in 1634, that was a typical blend of reasoned Renaissance science and lingering medieval supernaturalism. The main character of the piece, a young Icelander named Duracotus, travels to the moon with the aid of his mother, an accomplished witch–an arrangement not unfamiliar to Kepler since his own mother was tried, though not convicted, of witchcraft.

Kepler was inspired to think about a trip to the moon, and of finding life there, by the recent discoveries of Galileo. In *The Starry Messenger*, Galileo had gone beyond merely noting that the moon was "not unlike the face of the Earth," to speculating that its dark regions might be seas and its bright parts land and that "the moon has its own atmosphere." Although within a few years he came to doubt these conclusions, they were given wide credence by others. Kepler could claim to be the first to hypothesize about extraterrestrial life based on instrument-gathered data. He wrote to Galileo suggesting that one of the large features he (Galileo) had seen on the lunar surface might have been excavated by intelligent inhabitants who "make their homes in numerous caves hewn out of that circular embankment." Galileo wouldn't be drawn on this except to say that if there were lunar life, it would be "extremely diverse and beyond all our imaginings."

Galileo's lunar observations stirred the creative juices of several other seventeenth-century intellectuals who wrote about space travel, both in fantasy and theory. The English churchman Francis Godwin, who served as bishop of Hereford, authored what's arguably the first science fiction story in the English language. In *The Man in the Moone*, published in 1638, five years after his death, Godwin conveys his protagonist, Domingo Gonzales, to the moon in a craft towed by domestic geese. (Gonzales had intended a less ambitious flight but discovers en route that the birds are in the habit of migrating a little further than ornithologists had supposed.)

A couple of years later, another English bishop and the brother-in-law of Oliver Cromwell, John Wilkins, took up the subject of moon travel in the third edition of his *Discovery of a New World in the Moone*. He even suggested that colonies might be established there, a proposal that, not surprisingly, drew derogatory comments from foreign writers about British imperialism. Among those who poured scorn on his ideas was the orator and sermonizer Robert South, who suggested that Wilkins had ambitions to obtain a bishopric on the moon.

Cyrano de Bergerac, the French author and prolific duelist–most of his duels fought on account of his extraordinarily large nose–also weighed in with a combination of bizarre and clever ideas. His *L'Autre Monde: ou les Estats et Empires de la Lune* (The Other World: or the States and Empires of the Moon) and *Estats et Empires du Soleil* (The States and Empires of the Sun), both published posthumously in 1656, included the reappearance of Godwin's Domingo Gonzales and probably helped inspire Swift's *Gulliver's Travels* and Voltaire's *Micromegas*. Six of the methods of space propulsion Cyrano outlined in *L'Autre Monde* were completely off the wall, including the use of vials of rising dew and a smoke-filled, balloonlike globe. A seventh technique was remarkably prescient, however: an exploratory vehicle fitted with solid-fuel (gunpowder) rockets.

A Cannon to the Cosmos

A decade later, Isaac Newton began teasing out the physics upon which actual excursions into space would depend, and in *Principia*, he offered the first recognizable description of an artificial satellite. Newton imagined a stone being thrown straight out from the top of a high mountain: "[T]he greater the velocity with which [the stone] is projected, the farther it goes before it falls to the earth. We may therefore suppose the velocity to be so increased, that it would describe an arc of 1, 2, 5, 10, 100, 1000 miles before it arrived at earth, till at last, exceeding the limits of the earth, it should pass into space without touching."

Say the mountain was a very unearthly 200 miles high, forty times taller than Everest, so that it poked above the last vestiges of atmosphere. A stone hurled horizontally at about 18,000 miles per hour from the summit of this lofty peak would travel just fast enough that it would fall at the same rate at which the Earth curved away beneath it. Therefore, assuming there was no residual effect of air resistance, which in fact there would be, it would remain at that height in a circular orbit. If the stone went any slower, it would fall faster than the Earth could curve away from it and so would eventually descend to the surface. If it went faster, its orbit would stretch out into an ellipse. If it went fast enough, it would break orbit altogether and head away from the Earth permanently on an open trajectory.

Escaping Earth's gravitational pull all in one go–without the need for further propulsion–demands an outbound starting speed at the planet's surface of around 25,000 miles per hour. One of the first people to think seriously how this might be achieved, and to build an engaging story around it, was Jules Verne in his novels *From the Earth to the Moon* (1865) and its sequel *Around the Moon* (1870). Verne wrote of a giant cannon, the Columbiad, that could fire astronauts clear into space and away from their home planet. The 900-foot-long Columbiad, with a bore of nine feet, was sunk vertically into the ground in Florida not far from today's real complex of launch facilities at Cape Canaveral. The first 200 feet of the barrel was packed with 122 tons of guncotton, which, when ignited, was enough to blast an aluminum capsule containing three men and two dogs to a speed of just over ten miles per second. Even after being slowed by its passage through Earth's atmosphere, the shell was still traveling at nearly seven miles a second–sufficient to reach the moon.

Although Verne used bona fide engineering calculations in the design of his cannon and lunar projectile, he was wildly optimistic about the crew's chances of survival. In reaching Earth-escape velocity inside the barrel of a gun, the passengers would have been subjected to instantly lethal levels of acceleration. The ingenious system of hydraulic shock absorbers devised for the floor of the projectile would have done nothing to save the fictional occupants, only one of which, the dog Satellite, expires during launch. Verne also took a liberty with the crew's means of disposing of the dead animal: opening a hatch in the capsule "with the utmost care and dispatch, so as to lose as little as possible of the internal air."

Aside from these technical implausibilities, Verne's biggest scientific howler was his treatment of weightlessness. He thought this effect would occur only at the so-called neutral point, where the pulls of gravity of Earth and the moon exactly cancel. Consequently, he allowed his crew only an hour or so of near-weightless conditions during their entire flight. The truth is a bit different. Although the astronauts would be under the influence of both Earth's and the moon's gravity throughout their journey, they would be in a weight-free state most of the time, as the crews of the Apollo missions to the moon can testify.

Weightlessness, or zero-gravity, happens not where there's an *absence* of gravity–no such place exists in the universe–but where gravity *is the only force acting*. The screaming passengers in a freely plunging theme park ride are (briefly) weightless, as is the crew of an orbiting shuttle or space station. In both cases, vehicle and contents travel together along the same path and fall without restraint under the influence of gravity alone. The term *zero-gravity* is especially confusing as it seems to suggest there's no gravity at all. Yet, the occupants of a spacecraft orbiting Earth are obviously not beyond the reach of Earth's gravity because it's this very force that, like a string tied to a whirling stone, prevents the craft from flying off at a tangent. Likewise, if the sun's gravity were to be suddenly switched off, Earth would hurtle away on a straight-line course in the direction it was heading–and with the speed it had at the instant its gravitational shackles were removed.

At an altitude of 200 miles, according to the inverse-square formula figured out by Hooke and Newton, gravity is only about 10 percent weaker than it is at the Earth's surface. This means that if a 200-mile tower were built, the occupants of the top-floor penthouse would still weigh nine-tenths of what they did at sea level. Those residents wouldn't be weightless, as astronauts would be if they were orbiting at the same height, because they'd have a floor, ultimately fixed to the ground, to stand on. Gravity, therefore, wouldn't be the only force in the picture–there would also be the upward reaction of the floor against their feet. In the same way, a skydiver experiences weightlessness for only a few seconds before his downward speed gets high enough for air resistance, or drag, to become an important force in addition to gravity.

Rrrocket!

The great Victorian battle to escape Earth's clutches was fought on many fronts. Verne's mighty cannon was just one of an extraordinary assortment of fictional schemes devised for conveying the well-heeled ladies and gentlemen of this period to the moon and beyond. In *A Honeymoon in Space*, George Griffith, a popular rival to H. G. Wells in his heyday, told how the eccentric earl of Redgrave and his American wife celebrated their marriage with a postnuptial tour of the moon and planets. Their propeller-driven vessel, like many imagined spacecraft of this era, looked more like a glorified airship than anything designed to reach escape velocity in a hurry or to cross the vacuum between worlds. More esoteric as a form of propulsion was "apergy," an antigravity principle used to power a spacecraft from Earth to Mars in Percy Greg's *Across the Zodiac* (1880) and borrowed for the same purpose by John Jacob Astor in *A Journey in Other Worlds* (1894). Most famously, in *The First Men in the Moon* (1901), H. G. Wells used movable shutters made of Cavorite, an imagined metal that shields against gravity, to navigate a spacecraft to the moon.

Meanwhile, back in the real world, a handful of venturesome scientists and engineers were making good progress in their efforts to show that space travel was a practical possibility. To accelerate steadily out of Earth's gravity, these visionaries realized that a powerful, continuous thrust was needed–the kind of thrust that only a rocket could deliver. Rockets work in a very simple way: you throw something out the back (the exhaust) and you get a same-sized forward shove (thrust) in return. Newton's third law of motion, "To every action, there's an equal and opposite reaction," is the rocketeer's touchstone.

Possibly the first person in history to suggest using rocket power as a means of traveling to and through space was a man slated for execution. Nikolai Kibalchich, a Russian medical student, journalist, and revolutionary, had attempted to kill Czar Alexander II on several occasions. Finally, along with some accomplices, he succeeded in assassinating the ruler on March 13, 1881, and on April 3, at the age of twenty-seven, was himself put to death. While in jail awaiting his fate, Kibalchich wrote some remarkable notes illustrating the principle of space propulsion. In them, he described a means of accelerating a platform by setting off gunpowder cartridges in a rocket chamber. Changing the direction of the rocket's axis, he realized, would alter the vehicle's flight path. "I am writing this project in prison," he explained, "a few days before my death. I believe in the practicability of my idea and this faith supports me in my desperate plight."

Kibalchich's work was closely followed by that of his compatriot Konstantin Tsiolkovsky, who produced some amazingly advanced ideas around the turn of the twentieth century. Inspired in his youth by reading Verne, Tsiolkovsky derived the so-called rocket equation, a formula that underpins how all rockets perform. He was also the first to show, mathematically, the advantage of using multistage rockets–vehicles consisting of more than one set of fuel tanks and rocket engines that can be discarded, stage by stage, as the fuel is progressively exhausted. Staging, he realized, was the key to reaching orbit and to escaping from Earth's gravity pull altogether. He pointed out:

> If a single-stage rocket is to attain cosmic velocity it must carry an immense store of fuel. Thus, to reach the first cosmic velocity [his term for the speed needed to enter Earth's orbit], . . . the weight of fuel must exceed that of the whole rocket (payload included) by at least four times. . . . The stage principle, on the other hand, enables us either to obtain high cosmic velocities, or to employ comparatively small amounts of propellant components.

Tsiolkovsky also thought deeply about the biological problems of high- and low-gravity forces in space travel. He proposed immersing astronauts in water to reduce the effects of acceleration at takeoff and wrote about spacesuits and even zero-gravity showers–technology that would be brought to bear in his vision for colonizing the solar system.

His German contemporary Hermann Ganswindt introduced another futuristic idea for long-duration spaceflight: artificial gravity. Anyone familiar with the carnival ride in which people seem to stick to the vertical wall of a rotating cylinder will grasp the principle. The reaction force a person or object experiences against the inside of a spinning tube is very much like gravity. Make the tube big enough and its rate of spin just right, and the effect is indistinguishable from natural gravity on Earth–an environment much healthier for humans to live in for long periods than the weightlessness of normal space travel. Ganswindt proposed creating artificial gravity aboard an interplanetary spaceship by rotating a section of the craft at the appropriate speed. In 1928, the Austrian engineer Herman Potocnik, better known by his pen name Hermann Noordung, put forward a similar scheme involving a space station.[24] Potocnik's design envisaged a rotating, 100-foot-diameter, doughnut-shaped structure to serve as living quarters, with an airlock at its hub, a solar power generator attached to one end of the central hub, and an astronomical observatory. This concept of a lazily spinning, wheel-shaped space station was taken further by Wernher von Braun in the 1950s and is probably familiar in most people's minds from its spectacular depiction in the 1968 film *2001: A Space Odyssey* to the tune of Strauss's "Blue Danube Waltz."

From fiction to theory to experimentation, the means of leaving behind and living beyond Earth were explored. Practical rocketry developed rapidly from the 1920s onward in the hands of engineers such as Robert Goddard in the United States and members of the German *Verein für Raumschiffahrt* (VfR) (Society for Space Travel), formed in 1927 to put into action some of the theoretical ideas set out in Hermann Oberth's 1923 book *Die Rakete zu den Planetenräumen* (The Rocket into Interplanetary Space).[25] Both Goddard and the VfR experimented with early versions of liquid-fueled rockets—the technology that, within a few decades, would help loft humans and their machines to the edge of space.

A prominent figure in the VfR was von Braun, who'd earlier assisted Oberth in his efforts to build a small rocket as a publicity stunt for the movie *Frau im Mond* (Woman in the Moon), for which Oberth served as technical adviser. (An explosion during a bench-top test of the rocket cost Oberth the sight in one eye.) Von Braun was eventually drawn into the Nazi advanced weapons program and played a central role in developing the first large, liquid-propellant rocket, the V2, used to deadly effect in bombing London and other European cities during World War II. After the war, von Braun and more than a hundred of his colleagues, together with a large stock of V2s and V2 components, were spirited out of Germany to the United States in a scheme known as Operation Paperclip.

Once on American soil, the German team began military missile work for their new employers and later formed the backbone of the U.S. space program. Von Braun himself became the director of the National Aeronautics and Space Administration's (NASA's) Marshall Space Flight Center and oversaw the design and construction of the colossal Saturn V rocket that would carry men to the moon. A similar but smaller harvesting of German expertise supplied the nucleus of the Soviet Union's post-War effort to send people, equipment, and, if necessary, nuclear warheads on orbital and suborbital flights.

The conquest of space began in earnest on October 4, 1957, with the launch of Sputnik 1, the first human-made object to defy the old adage "What goes up must come down." At least, it defied it for several months before its orbit decayed and it incinerated in the atmosphere. Further triumphs in the struggle to shake off Earth's gravitational clutches came thick and fast. Several more Soviet and American satellites were lobbed into orbit before Luna 1 became the first probe to achieve escape velocity and hurtle past the moon, in January 1959. Two years later, mankind began to follow its creations to the high frontier when Yuri Gagarin circled once around his home planet on a flight lasting 108 minutes.

Gee Whiz

Never before had there been a test in situ of how the human body would cope with the rigors of spaceflight–in particular, with the stresses arising from the high *g*-forces of launch and reentry and the strange new condition of weightlessness. But the physiological effects of being slammed by forces equivalent to many times Earth's normal gravity had been under medical scrutiny since the late 1940s. In an almost unbelievable series of experiments conducted mostly on himself, the American physician John Stapp, one of the great pioneers of aerospace medicine, explored the limits of acceleration tolerance during terrifying rides aboard rocket-powered sleds.

In 1946 Stapp joined the Aero Medical Lab at Wright Field, Ohio, and served as flight surgeon to the legendary Chuck Yeager when he broke the sound barrier. Stapp became convinced that a significant pattern lay behind the way some airmen died and others survived seemingly equally violent crashes. To solve the mystery, he began a program of experiments using a rocket sled christened *Gee Whiz* at Muroc Air Force Base (later renamed Edwards Air Force Base) in California. Stapp developed a special safety harness and perfected it during thirty-two sled runs that involved a dummy called Oscar Eight-Ball. Finally, he was ready to try out the restraint on a human guinea pig–himself. Strapped into the sled facing rearward, and having refused an anesthetic because he wanted to study his reactions firsthand, Stapp was hurled to 90 miles per hour in less than a second and then was crushed against the seat back, straining several muscles as the sled ground to an almost instantaneous halt. Within a year, Stapp had made sled runs at up to 150 mph, stopping in as little as 19 feet and experiencing as much as 35 *g*, or thirty-five times the normal acceleration due to gravity at the Earth's surface. For comparison, the maximum gravity that a space shuttle astronaut pulls on her trip to orbit is only about 3 *g*. Although in the process of his radical experiments Stapp suffered headaches, concussions, a fractured rib and wrist, and a hemorrhaged retina, he proved that the human body could withstand such punishment and even remain conscious throughout the ordeal.

When Stapp's commanding officer learned he'd acted as his own guinea pig, he ordered the sled runs to stop, fearing he'd miss out on promotion if Stapp were killed. Stapp, however, secretly continued the tests by using chimpanzees. He found that when the apes were strapped in correctly, they survived forces many times those experienced in most plane crashes. From this he concluded that crash survival doesn't depend on a body's ability to withstand the high forces involved but rather on its ability to avoid the mangling effects of the vehicle.

To back up this idea, Stapp unofficially resumed his human tests–putting himself, as usual, first in the firing line. Over the next four years, he lost fillings, cracked more ribs, and fractured his wrist again. In 1949 he was involved in the birth of Murphy's Law. Stapp's harness held sixteen sensors to measure the g-force on different parts of the body. There were exactly two ways each sensor could be installed, and it fell upon a certain Captain Murphy to make the connections. Before a run in which Stapp was particularly badly shaken up, Murphy managed to wire up every sensor the wrong way, with the result that when Stapp staggered off the rocket sled with bloodshot eyes and bleeding sores, all he had to show for it was a bunch of zero sensor readings. Known for his razor-sharp wit, Stapp quipped, "If there are two or more ways to do something and one of those results in a catastrophe, then someone will do it that way."

The advent of supersonic flight and the need to bail out at very high speed demanded more extreme experiments. Transferred to head the aeromedical field lab at Holloman Air Force Base in New Mexico, Stapp built a much faster sled called the Sonic Wind No. 1. He did some initial testing with dummies, but in March 1954 he again put himself forward as the subject. In his first ride on the new sled, Stapp clocked 421 mph–a new land-speed record. On December 10 he took the sled chair for his final and most extreme ride. His wrists were tied together in front of him because a flapping limb would be torn away in the ferocious air stream. His major concern was that the rapid deceleration might blind him. Earlier, he had "practiced dressing and undressing with the lights out so if I was blinded I wouldn't be helpless." At the end of the countdown, Stapp was shot to a mind-boggling 623 mph in five seconds and back to rest in just over another second. Subjected to 40 *g*, he temporarily blacked out, and his eyeballs bulged from their sockets. He was rushed to the hospital; his eyesight gradually returned, and a checkup revealed he had suffered no major injury. An hour later, he was eating lunch.

Stapp subsequently helped run tests on human and animal subjects in the giant Johnsville Centrifuge–the nightmare whirligig built by the U.S. Navy and used in selecting and training the Mercury astronauts. At the other end of the gravity spectrum, the Gimbal Rig set up at the Lewis Research Center in Cleveland, Ohio, was used to give future space travelers a feel for the disorienting effects of zero-*g*. In later years, astronauts would be able to prepare for performing tasks in weightless conditions in a giant water tank known as the Neutral Buoyancy Simulator at NASA's Marshall Space Flight Center in Huntsville, Alabama, or get a taste of genuine zero-g before venturing into space by flying aboard the infamous "vomit comet"–the space agency's converted KC-135 transport plane. By climbing and then swooping in a parabolic arc while at the same time using its engines to compensate for air resistance, the plane provides twenty to thirty seconds of true free fall. After a few dozen such dives, a would-be astronaut arrives at a new appreciation for the terms *weightless* and *space motion sickness*.

Weight for Me

In Gagarin's wake, other men and women ventured into space for trips lasting many hours or days. As space stations were built–Salyut, Skylab, Mir, and, most recently, the International Space Station–people were able to remain in orbit for months at a stretch, the current record being 438 days by Valeri Polyakov aboard Mir. Gradually, data began to accumulate that such long exposure to a weight-free environment has some disturbing effects on the human body.

Muscles, including that most essential muscle, the heart, progressively weaken and shrink without gravitational loading. Bones demineralize. Relieved of the normal stresses produced by gravity, bones start to lose calcium and other minerals faster than they can replace them. This isn't a problem on short missions, such as those of the space shuttle. But some of the bones of astronauts and cosmonauts who've spent months aboard space stations have been found to have lost up to one-fifth of their mineral content, increasing the risk of fractures. Not all bones lose minerals at the same rate in space. Bones in the upper body, for instance, don't seem to be affected at all, whereas the weight-bearing bones in the legs and lower back lose more than their fair share.

Plenty of exercise is an essential part of the daily routine aboard space stations in an effort to combat both bone demineralization and muscle degeneration. But the forces needed to prevent deterioration of the weight-bearing bones seem to be roughly equal to the body weight of the individual–not surprisingly, since most people gain and lose bone minerals at the same rate doing normal activities on Earth. This makes it hard to design exercise equipment that can produce an effective level of force in a way convenient for astronauts to use. Various methods have been proposed and tried, including bungees, springs, bicycles, and treadmills, but attaching the load-bearing part of the exercise equipment to the body in an acceptable way has been a stumbling block. Straps tend to cut into the shoulders and hips, and the human shoulder isn't designed to carry body-weight loads for long periods or during intensive exercise.

For missions to the planets, which may last two, three, or

more years, the only foreseeable way to avoid potentially life-threatening medical problems is artificial gravity. A habitable portion of a long-haul crewed spacecraft will need to be able to rotate to substitute for at least a sizable fraction of the gravitational loading that would be present on Earth. This poses a major challenge because the artificial gravity section has to be big enough in diameter that it doesn't have to rotate so quickly as to produce nausea. Some compromise seems inevitable; perhaps the best that can be practically managed is an artificial gravity that's only a fraction of what would be normal on Earth. Alternatives to a revolving cylindrical section of a spacecraft have been suggested, including a habitable artificial gravity module at the end of a long tether that sweeps around the main vessel in a wide circle.

When permanent colonies are established on the moon and Mars, it seems inevitable that, at some point, babies will be born on these low-gravity worlds. Then the question will arise as to what to do with these "alien" children: return them to Earth to grow up in the gravitational regime for which their genes have evolved or allow them to remain on their birth world so that their bodies become permanently adapted to life at low gravity? In the latter case, there's the issue of how much gravity an off-world child would need in order to grow into a healthy adult; perhaps Mars with its surface gravity of 0.38 Earth value would provide enough loading for developing muscles and bones, but the moon, with only 0.17 Earth's surface gravity, would lead to fatal weaknesses in the musculoskeletal system. A life in low gravity would certainly have its advantages, including less sagging skin and flesh in later years. But those who'd spent all their childhood in less than half the gravity pull of Earth would have to resign themselves to one sobering fact: they would never be able to return to the cradle of mankind.

Grav-ET

As for life that may have evolved independently on planets with higher or lower surface gravities than Earth, there has been no shortage of speculation. The prodigious strength-to-mass ratio of some terrestrial organisms–the rhinoceros beetle, for example, can manipulate objects with up to 850 times its own body mass–suggests that even complex mobile creatures might thrive on worlds where the gravity pull is a lot higher than Earth's, providing that other basic biological requirements weren't compromised. Some scientists and science fiction authors have gone further and entertained the possibilities for life where the surface attraction is far beyond the normal planetary range. The American science fiction writer Hal Clement (born Harry Clement Stubbs), who has degrees in astronomy and chemistry, speculated in detail about the intelligent lifeforms that might evolve on a world of extremely high gravity. His novels *A Mission of Gravity* (1954), *Close to Critical* (1964), and *Starlight* (1971) explore the biology and physics of the giant, rapidly spinning planet Mesklin, the surface gravity of which varies from three times (at the equator) to seven hundred times (at the poles) that of Earth.

Even more extraordinary would be the conditions facing organisms that resided on the surface of a neutron star–the bizarre biological environment hypothesized by Frank Drake, the SETI (Search for Extraterrestrial Intelligence) pioneer, and Robert Forward, a space propulsion expert and fiction writer. A neutron star is the supercollapsed remains of a star, the outer parts of which have been blasted away in a supernova explosion. With around a couple of solar masses of material crammed into a ball less than twelve miles across, a typical neutron star is so dense that a teaspoonful of it would outweigh the human race. The gravitational pull at its surface would be about seventy billion times stronger than that on Earth. With tongue slightly in cheek, Drake speculated that life might exist on its solid surface, which is more like that of a planet than a normal star. The creatures he imagined were submicroscopic and made of tightly packed nuclei, rather than ordinary atoms, bound together as "nuclear molecules." Whether such bizarre molecules could exist and combine in ways complex enough to give rise to life isn't known. If neutron star creatures did exist, however, they would almost certainly live at lightning speed. Nuclear reactions happen much faster than the chemical variety, so that any life-forms on a neutron star would probably evolve and live their lives a million times more quickly than human beings. Forward later developed and elaborated these ideas in two novels, *Dragon's Egg* and *Starquake.*

At the opposite extreme, the British astrophysicist Fred Hoyle mused in his 1957 novel *The Black Cloud* how total freedom from planetary gravity in interstellar space might allow intelligence to evolve to a much higher order. Hoyle describes the arrival near Earth of a small interstellar cloud that can think and move of its own accord. A living organism, half a billion years old, as big as the orbit of Venus, and as massive as Jupiter, the Black Cloud has a brain that consists of complex networks of molecules that can be increased in number and specialization as the creature desires. Once it learns that the third planet of the Sun is inhabited by an intelligent race, it makes contact and begins to reveal some extraordinary facts about intelligence in the universe:

> [I]t is most unusual to find animals with technical skills inhabiting planets, which are in the nature of extreme outposts of life. . . . Living on the surface of a solid body, you are exposed to a strong gravitational force. This greatly

> limits the size to which your animals can grow and hence limits the scope of your neurological activity. It forces you to possess muscular structures to promote movements, and . . . to carry protective armor. . . . [Y]our very largest animals have been mostly bone and muscle with very little brain. . . . By and large, one only expects intelligent life to exist in a diffuse gaseous medium.

Fantastic though such a creature might seem, Hoyle writes in his preface, "There is very little here that could not conceivably happen."

Gunning for Space

Other worlds and other life beckon, but the hardest part of space travel remains getting and staying just a few hundred miles above the ground. Once safely in low Earth orbit, it's pretty straightforward to redirect a spacecraft into a different or higher orbit or to dispatch it away from our planet altogether on a mission into deep space.

Multistage rockets are the only way at present to get payloads into orbit. Other methods, however, are waiting in the wings. The trouble with rockets is that they have to lift not only their own weight but also the weight of their fuel and the material used to burn that fuel, known as the oxidizer. A space cannon, like Jules Verne's Columbiad, gives far more bang per buck than a rocket because all the fuel is contained within the gun barrel and doesn't have to be carried along for the ride. But launching by cannon does have its drawbacks. The payload has to be skinny enough to fit snugly into a gun barrel and sturdy enough to withstand the huge accelerations of launch, which can easily top 10,000 *g*. Also, to get an object into orbit, even if it's been given a high enough initial speed, takes some kind of onboard propulsion to nudge the object gradually onto a sideways trajectory so that, at orbital height, it's moving more or less parallel with the ground.

Big, mean-looking cannons have been around for a while. A generation before the first test flights of the V-2, the "Paris Gun" of World War I blasted shells to impressive new altitude and speed records. Also known as the Wilhelm Geschuetz, after Kaiser Wilhelm II, this early supercannon was used by the German army to bombard the French capital from the woods of Crepy. It was a weapon like no other, capable of hurling a 207-pound shell to a range of 80 miles and a maximum altitude of 25 miles–the greatest height reached by a human-made projectile until the first successful V-2 flight test in October 1942. At the start of its 170-second trajectory, each shell from the Paris Gun was traveling at a mile per second, or almost five times the speed of sound.

The gun itself, which weighed 256 tons and was mounted on rails, had a 92-foot-long, 8.3-inch-caliber rifled barrel with a 20-foot-long smoothbore extension. After 65 shells had been fired, each of progressively larger caliber to allow for wear, the barrel was rebored to a caliber of 9.4 inches. The German goal of this extraordinary device wasn't to destroy France–it was far too inaccurate for that–but to erode the morale of the Parisians. From March through August 1918, three of the guns fired 351 shells at Paris from the woods of Crepy, killing 256 and wounding 620.

In the 1920s members of the German VfR amused themselves by redesigning Jules Verne's lunar cannon. The rocket pioneers Max Valier and Hermann Oberth came up with a revision of the gun that would correct Verne's technical mistakes and make it viable for shooting at the moon. The projectile would be made of tungsten steel, practically solid, with a diameter of almost 4 feet and a length of 24 feet. To avoid air compression in the 3,000-foot-long barrel during acceleration, Valier and Oberth suggested that the barrel be pumped out to create a near vacuum and a metal seal placed over the end. Any residual air pressure would be enough to blast this seal aside before the shell left the gun. To minimize drag in getting through the atmosphere, they proposed sinking the cannon into a mountainside at an altitude of at least 16,000 feet, which would put the mouth of the gun above most of Earth's atmosphere.

By the 1950s it was obvious that the rocket was going to become the chief means of getting into space for the foreseeable future, but that didn't stop others from pushing ahead with alternative schemes of their own. The Canadian engineer Gerald Bull began a lifelong effort to develop a supercannon for cheap access to the upper atmosphere and low Earth orbit. The High Altitude Research Project (HARP), carried out in the 1960s by Bull and his colleagues from McGill University in Montreal, showed that a suborbital cannon could be cost-effective for studying the atmosphere at heights of 30 to 80 miles and had the potential to launch vast numbers of minisatellites each year in all kinds of weather. The HARP projectiles were cylindrical, finned missiles, eight inches wide and five and a half feet long, with masses of 175 to 475 pounds, called martlets, from an old name for the martin bird that appears on McGill's shield. The gun that propelled them was built from two ex–U.S. Navy 16-inch-caliber cannons joined end to end. Located on the island of Barbados, the cannon fired almost vertically out over the Atlantic.

Inside the barrel of the cannon, a martlet was surrounded by a machined wooden casing known as a sabot, which traveled up the 52-foot-long barrel at launch and then split apart as the martlet headed upward at about five miles per second, having undergone an acceleration of 25,000 *g*. Each shot produced a plume of fire rising many hundreds of feet into the air and a huge explosion that could be heard all over the island. The martlets carried payloads of metal chaff, chemical smoke, or meteorological balloons and were fitted with telemetry antennas for tracking their flight. By the end of 1965, HARP had fired more than a hundred martlets to heights of over 50 miles. On November 19, 1966, the Army Ballistics Research Laboratory used a HARP gun to launch a 185-pound martlet to an altitude of 111 miles–a world record for a fired projectile that still stands.

Further work on the supercannon was carried out by the Lawrence Livermore National Laboratory in California, as part of its Super High Altitude Research Project (SHARP). The launcher in this case was a light-gas gun, funded by the Strategic Defense ("Star Wars") Initiative as a possible antimissile defense weapon. It consisted of a 269-foot-long, 14-inch-caliber pump tube and a 154-foot-long, 4-inch-caliber gun barrel, in an L-shaped arrangement. This setup was chosen to avoid one of the main problems with a space cannon: the projectile can't outrun the gas molecules that push it along the barrel. The speed of these molecules is higher the smaller the mass of the gas molecule. The lightest and best gas for the job is hydrogen, but hydrogen isn't produced as a product of any explosive mixture.

The solution in SHARP was to use a gun with two connected barrels–an auxiliary one and a main one in which the payload is accelerated. SHARP went into operation in 1992 and demonstrated velocities of 1.6 miles per second (about eight times the speed of sound) with 11-pound projectiles fired horizontally. Impressive as this sounds, it falls well short of what is needed to fire projectiles into space–at least 24 times the speed of sound for a low-altitude circular orbit–even if the barrel were pointed upward. Moreover, the $1 billion needed to fund space launch tests never materialized.

In 1996 project leader John Hunter founded the Jules Verne Launcher (JVL) Company to develop the concept commercially. After some prototype work, the company planned a full-scale gun that would have been bored into an Alaskan mountain for launches into high-inclination orbits. It would have been able to fire 11,000-pound projectiles, each 5.6 feet in diameter and 30 feet long, with a muzzle velocity of 4.3 miles per second. Following burn of the rocket motor aboard the projectile, a net payload of 7,300 pounds would have been placed into low Earth orbit. Experience with SHARP suggested that the space gun could have been fired up to once a day. Thus, a single gun could have placed over 1,000 tons a year into orbit at a cost per pound one-twentieth that of conventional rocket launchers. JVL continued in business until the late 1990s, but no investors came forward to finance the multibillion-dollar development cost.

Sky Lift

An even cheaper and more convenient way of reaching Earth orbit was the 1960 brainchild of Russian engineer Yuri Artsutanov. Others, including John Isaacs and his colleagues at the Scripps Institute of Oceanography and Jerome Pearson of NASA's Ames Research Center, independently worked on the concept, but it went largely unnoticed until 1979, when Arthur C. Clarke made it the centerpiece of his novel *The Fountains of Paradise*.

The space elevator is known by a variety of other names: heavenly funicular (Artsutanov's original description), orbital tower, beanstalk, and sky hook. It is basically a cable stretching from a point on Earth's surface to another point 29,000 miles directly above it. Passengers and cargo would ride up and down the cable in a manner similar to a conventional elevator–or, more accurately, a cable car–traveling at a fraction of escape velocity. That would slash the cost of putting payloads into orbit to as little as 67 cents per pound, compared with $10,000 per pound on a rocket. Moreover, you wouldn't have to be a superfit astronaut to make the trip, thus opening up space to the enterprising masses.

To understand how the space elevator works, think of a satellite. The time it takes to orbit Earth is determined by the strength of gravity, and this varies with altitude: low-flying satellites orbit quickly, distant ones much more slowly. At the special altitude of 16,232 miles, a satellite takes exactly one day to complete one orbit. If its orbit is aligned with the equator, a satellite at this distance will hover over the same point on Earth's surface as the two turn in celestial tandem. Satellites parked in such an orbit arc said to be geostationary. Now imagine stretching the satellite in toward the Earth, and at the same time, outward into space, so that its center of mass remains in geostationary orbit. Those parts of the satellite closer to Earth will be moving more slowly than necessary to maintain a stable orbit, and so will start to feel gravity's pull. In contrast, the parts further away will be moving too quickly for their distance and so, like a stone in a sling, will try to move further afield. The result: tension. The satellite becomes a taut cable in orbit. Now carry this thought experiment to its logical conclusion, where the satellite's innermost point strikes ground zero–or, more likely, connects to a tall tower.

The result is a continuous structure stretching all the way from the equator into space. At the Earth end is the base station, a massive complex with all the trappings of a major international airport–hotels, restaurants, duty-free shops, and the like. Looming above this is the launch structure, maybe tens of miles tall. Then comes the cable, all 29,000 miles of it, uninterrupted except for a space station at the geostationary point.

This would serve as the structure's center of mass as well as housing labs, a business park, and a zero-gravity resort. Further out lies a counterweight, possibly a minor asteroid tethered to the end of the cable.

It sounds too fantastic to be true, but only a few technological hurdles stand in our way. By far the greatest of these is the cable itself. The sheer weight of the structure dangling from geostationary orbit would place extraordinary demands on the material used to make it. What sort of stuff has the tensile strength needed to support its own weight over such a length? For a cable of practical dimensions, widest at geostationary orbit, where the tension is highest, tapering to a minimum diameter of four inches at the either end, estimates are that it would need to be made of something thirty times stronger than steel or seventeen times stronger than Kevlar. One possibility is carbon in the form of so-called nanotubes: tiny, hollow cylinders made from sheets of hexagonally arranged carbon atoms. At present nanotubes are extremely expensive and can be fabricated only in short lengths. It seems likely that production costs will fall dramatically in the future, however, and that some way will be found to bind nanotubes into a composite material like fiberglass.

The Currents of Space

Once a spacecraft leaves the safe harbor of Earth orbit, it's at the mercy of the gravitational eddies and currents that swirl around the solar system. The sun and planets, together with their moons, give rise to ever-shifting patterns of gravitational force that can suck in a passing probe or hurl it away on a path to oblivion. Careful planning and savvy calculations, on the other hand, allow space missions to take full advantage of the complex play of gravity.

With the advent of satellites and space probes, accurate calculations of trajectories, taking into account the gravitational pull of Earth, the sun, the moon, and the planets, became crucial. These calculations, only possible with high-speed computers, were one of the motivating factors behind the development of the integrated circuit. As time went on, spaceflight planners and engineers learned how to take advantage of the gravitational wells and streams in the solar system to speed probes on their way or allow them to reach their destinations while carrying less fuel.

The most effective economizing trick is gravity assist, sometimes called the slingshot effect, in which the gravitational field of a planet is used to change–usually increase–the velocity and alter the course of a vessel passing close by without the need to expend fuel. The inbound flight path is carefully chosen so that the spacecraft is whipped around the assisting body, being both sped up and deflected on a hyperbolic trajectory. At first sight, it may seem as if something has been gained for nothing. However, the additional speed of the spacecraft has been won at the planet's expense, which, as a result of the encounter, slows imperceptibly in its orbit and, as a result, moves fractionally closer to the sun.

One of the earliest and most dramatic applications of the technique came in 1970 when the world watched as NASA used a lunar gravity-assist to rescue the Apollo 13 astronauts after an onboard explosion had severely damaged their spacecraft en route to the moon. By using a relatively small amount of fuel to put the spacecraft onto a suitable trajectory, NASA engineers and the astronauts were able to use the moon's gravity to turn the ship around and send it back home. Numerous interplanetary missions, including those of Pioneers 10 and 11, Voyagers 1 and 2, Galileo, and Cassini, have also successfully used the method to hop from world to world or be accelerated toward their final goal.

More recently, a more subtle gravity exploitation trick has been brought to bear, which, like the slingshot effect, enables interplanetary missions to be flown with much smaller amounts of propellant and therefore at lower cost. First used by NASA in the 1980s to maneuver ICE (International Cometary Explorer), it involves tapping the gravitational properties of certain special points in the orbit of one body as it goes around another–the so-called unstable Lagrangian points, L1, L2, and L3. (A Lagrangian point is a location in space around a rotating two-body system, such as the Earth-moon or Earth-sun, where the pulls of the gravitating bodies combine to form a point at which a third body of negligible mass would be stationary relative to the other two.) Lagrangian points are named after the Italian-born French mathematician and astronomer Joseph Louis de Lagrange (1736–1813), who first showed their existence. There are five Lagrangian points in all, but three are unstable because the slightest disturbance to any object located at one of them causes the object to drift away permanently.

The basic idea is that a tiny nudge to a spacecraft at one of these points, costing very little fuel, can cause a surprisingly big change in the spacecraft's trajectory. Following NASA's initial trial-and-error approach with ICE, the mathematics of chaotic control began to be developed properly in 1990, starting with some work of Edward Ott, Celso Gregobi, and Jim Yorke of the University of Maryland.[27] Their method, known by their initials as the OGY technique, involves figuring out a sequence of small maneuvers that will give the desired overall effect. NASA exploited this more refined version of chaotic control on the Genesis mission to harvest samples of the solar wind. At the end of its two-and-a-half-year collection phase, Genesis didn't have enough fuel for a direct return to Earth. Instead, it was first sent on a long detour to the L2 point, outside Earth's orbit around the sun, from which it was brought back very economically to the Earth-moon L1 point and from there, by way of a few cheap chaotic orbits of the moon, into a stable Earth orbit. Finally, its cargo capsule was released for a (much harder than anticipated) landing on the salt flats of Utah in August 2003.

Even the most sophisticated of spacecraft maneuvers can be successfully calculated using the laws of physics as they were understood in the seventeenth century. Computers may be needed to solve the math, but the underlying science is that of Newton. When the Gravity-B probe rode toward Earth orbit on April 20, 2004, the flight of its Delta II rocket from Vandenberg Air Force Base, in southern California, would have held no mystery to the author of *Principia.* Yet Newton would have been fascinated by the spacecraft's mission: to test a theory of gravity that even his great genius could never have foreseen.

6

One of Our Planets Is Missing

The great tragedy of science—the slaying of a beautiful hypothesis by an ugly fact.

—THOMAS HUXLEY

The fall of every great theory is foreshadowed by some niggling problems or inconsistencies that refuse to go away. By the end of the nineteenth century, several unresolved questions hung over Newton's universal theory of gravitation, and some scientists were beginning to ask if it might not be time for a change, for a new vision of the way gravity works.

Predictions and Confirmations

Not that it was obvious to most people that the greatest theory known to science was in trouble. Far from it; the physics of Newton's *Principia* looked to have been a resounding success and one set to last. In the century and half since Newton had laid down his axioms of motion, they'd been refined and made more potent by mathematicians of the caliber of Leonhard Euler, Joseph Lagrange, William Hamilton, and Karl Jacobi.

Newton's view of gravity, meanwhile, had emerged victorious over its chief rival on the Continent–Descartes' vortex theory–even though, for a while, that cross-Channel contest had seemed close. In the end, French mathematicians were won over to Newton's strange ideas of action at a distance and inverse square behavior by breakthroughs achieved among their own ranks.

In 1743 the French mathematician Alexis Clairaut went to Lapland as part of an expedition led by Pierre Maupertuis to collect data to help determine the exact shape of the Earth. In his *Théorie de la Figure de la Terre* (Theory of the Shape of the Earth), Clairaut marshaled this data to confirm a belief shared by Newton and the Dutchman Christiaan Huygens that the world is squashed at its poles into the shape of an oblate spheroid. A couple of years later, he began work on another puzzle that had taxed Newton: the so-called three-body problem.

The goal of this problem–now known to be unreachable except in certain special cases–is to figure out the exact movements of three objects that are interacting with one another gravitationally. Focusing on the moon, the orbit of which is strongly affected by both the Earth and the sun, Clairaut reached a startling conclusion: Newton's theory of gravity didn't work; there was something wrong with the inverse square law. In an announcement to the Paris Academy of Sciences, he suggested that an extra term, involving one over the fourth power of the separation distance, needed to be included in Newton's formula. This was music to the ears of those who preferred Cartesian eddies any day to near-mystical Anglaise nonsense about forces acting across the interplanetary void. Euler, the greatest mathematician of his time, threw his weight behind Clairaut and briefly rejoined the vortex theory flock.

By the spring of 1748, however, Clairaut was ready to retract what he'd said. The difference between the observed motion of the moon's apogee–the point of greatest separation with Earth–and the one predicted by theory wasn't due to a glitch in Newton's law of gravity, he realized, but to errors coming from the approximations that had been made. Clairaut reported to the Paris Academy on May 17, 1749, that his theory was now in line with the inverse square law. The tide was beginning to turn back in Newton's favor, and soon it would be unstoppable.

Clairaut decided to apply his knowledge of the three-body problem to calculate the orbit of what eventually became known as Halley's comet and so predict the exact date of its return. When the comet reached perihelion only a month before the appointed date, Clairaut was the toast of his homeland, and Newton's views were no longer seriously challenged as the gravitational theory of choice.

It seemed now only a matter of time before fast-improving observations of the various planets, moons, and comets, harnessed to the math of Newtonian gravity, would be able to explain how all the objects in the solar system moved. Still, it was a struggle for theorists to keep up with their colleagues at the eyepiece who were taking full advantage of larger, better telescopes and other new astronomical gear. In 1781, William Herschel, a German-born amateur astronomer working from his private observatory in Bath, England, astonished everyone by discovering a new planet, twice as far from the sun as Saturn. Hovering at the threshold of naked-eye visibility, this world must have been glimpsed many times in history and prehistory; the earliest recorded sighting of it was in 1690 when John Flamsteed catalogued it as the star 34 Taurus. Herschel, however, was the first to recognize its true planetary status. He called it, inappropriately, Georgium Sidus (George's Star), after his patron King George III, but the name that eventually stuck was that of the Greek sky god, Uranus.

Close on the heels of this breakthrough, the French mathematician-astronomer Pierre Simon Laplace set about calculating an orbit for the new world. He found, as Clairaut had done in the case of Halley's comet, that when Uranus comes closest to massive Jupiter and Saturn, its orbit is perturbed–shifted a little from the perfect ellipse it would be if the sun were the only gravitating object present. With his compatriot Joseph Lagrange, Laplace calculated the perturbations of Uranus, together with those in the movements of the other planets, and showed that all were consistent with Newton's version of gravity.

Another Frenchman who made perturbations a central theme of his career was Urbain Leverrier, an assistant at the Paris Observatory and later its director. His research in this field, which began in 1838, led not only to a better knowledge of the masses of the planets and the scale of the solar system but also to predictions of two more planets around the sun. The first of these was subsequently confirmed; the other proved to be merely a ghost. But it was a ghost that would haunt Newton's universal theory of gravity into the next century–and ultimately into obsolescence.

Earlier, in 1821, Alexis Bouvard, also at the Paris Observatory, published a set of astronomical tables that gave the expected future positions of Uranus based on what was known about its orbit. A decade later, Uranus was a full 15 arc-seconds from where it was supposed to be–unignorably off course–and Bouvard had a pretty good idea why. The same thought occurred to the English mathematician John Couch Adams as revealed in notes he made while a student at Cambridge on July 3, 1841:

> Formed a design at the beginning of this week of investigating, as soon as possible after taking my degree, the irregularities in the motion of Uranus, which are as yet unaccounted for, in order to find whether they may be attributed to the action of an undiscovered planet beyond it; and, if possible, thence to determine the elements of its orbit approximately, which would lead probably to its discovery.

Four years later, in October 1845, Adams believed he had the missing world in his sights. He gave its anticipated path across the sky to Sir George Airy, the Astronomer Royal, but Airy, cautious and fastidious to a fault, sat on the results for a crucial nine months. In Germany, Friedrich Bessel, the director of the Königsberg Observatory, also had an eighth planet on his mind, but he died before he could complete his calculations.

Leverrier was more fortunate. At the urging of his superior, François Arago, the director of the Paris Observatory, he began to look closely at the wayward orbit of Uranus around the time Adams was completing his calculations. On June 1, 1846, Leverrier, too, came up with an X-marks-the-spot for a new outer planet, and after trying and failing initially to drum up interest in an observational search, he managed to persuade Johanne Galle, a student at the Berlin Observatory, to run a telescopic eye over his proposed coordinates. Heinrich d'Arrest, Galle's colleague at Berlin, suggested comparing a recently drawn celestial chart depicting the region of Leverrier's predicted location with the current appearance of the sky to try to spot any changes that might suggest an interloper. Sure enough, on September 23, 1846, the first night of searching, the new planet was found within one degree of where Leverrier said it would be and about ten degrees from Adams's predicted coordinates.

Given the age-old tradition of Anglo-French rivalry, claims and counterclaims about who knew what when were inevitable. James Challis, a British astronomer who belatedly started a hunt for the new world on July 29, 1846, at the encouragement of John Herschel, son of William Herschel, championed Adams's cause. In fact, Challis actually observed the planet discovered by Galle twice on August 4 but didn't identify it because of his casual approach–failing to compare it with what he had seen the night before. Herschel and Airy also supported Adams's claim of priority. But Arago saw it differently: "Mr. Adams does not have the right to appear in the history of the discovery of the planet Leverrier either with a detailed citation or even with the faintest allusion. In the eyes of all impartial men, this discovery will remain one of the most magnificent triumphs of theoretical astronomy, one of the glories of the Académie and one of the most beautiful distinctions of our country."

"Leverrier," of course, isn't the name by which the eighth planet of the sun became known, despite Leverrier's own unsubtle hint in writing a paper shortly after Galle's success in which he refers to Uranus as the planet Herschel, after William, its discoverer. Future generations would know the world beyond Uranus by the name first suggested by the German astronomer Johann Encke–Neptune.

It's ironic that the discovery of Neptune should mark the greatest triumph of Newton's theory of gravitation, whereas the failure to find another planet that Leverrier predicted should sow the seeds of its downfall. By the middle of the nineteenth century, all the worlds of the solar system appeared to be moving pretty much as they should according to Newtonian principles. Any anomalies were comfortably within the margins of observational error, and there were no obvious discrepancies between calculations based on the description of gravity in the *Principia* and how the planets and the rest of the sun's retinue moved, with one exception. The innermost planet, Mercury, seemed to be dancing to a slightly different tune from that of its neighbors.

The World That Wasn't

In Newton's theory, the orbit of a single planet going around the sun alone would be a perfect ellipse. Throw in some planetary companions, however, and that perfection is marred by the small gravitational tugs of these other worlds. One effect of these perturbations is to cause the point of closest approach to the sun–the perihelion–of a planet's orbit to revolve slowly around the sun. This so-called advance of the perihelion means that, over time, the orbit traces out the pattern of a rosette like that made by a child's Spirograph. In Mercury's case, Newton's theory showed that the advance of the perihelion ought to be 527 arc-seconds per century–by far the largest of any of the planets. But when Leverrier published his analysis of Mercury's motion in 1859, he pointed out that there was a discrepancy of 38 arc-seconds per century between the predicted advance and the observed value of 565 arc-seconds per century. By 1882 the discrepancy was known more accurately to be 43 arc-seconds per century.

All sorts of explanations were put forward to account for this. Perhaps Mercury had a moon, or Venus was 10 percent heavier than anyone had suspected, or the sun was more flattened at its poles than previously believed. Any of these ideas would have worked in theory, but, unfortunately, there wasn't a scrap of observational support for them. Leverrier favored a more exotic possibility–and began looking for evidence. Might there be, he wondered, another planet lying inside the orbit of Mercury, so close to the sun that it had never been noticed?

Leverrier's excitement was almost uncontainable when, on

December 22, 1859, he opened a letter from someone claiming to have seen just such a world. The message came from the country doctor and amateur astronomer Edmond Lescarbault, who had built himself a small observatory in the village of Orgères-en-Beauce, some 70 kilometers southwest of Paris. According to the letter, on March 26 of that year, using his modest 3-inch refracting telescope, Lescarbault had watched as a round black spot glided slowly across the face of the sun in an upward-slanting path. At first he took it to be a sunspot but then realized the motion was too quick for such a feature and, having observed the transit–the passage across the Sun's disk–of Mercury in 1845, he assumed that he was witnessing another such event, but of a previously undiscovered body.

In a fever of anticipation, Leverrier caught the next train to Orgères-en-Beauce, taking a colleague along as a witness, and arrived unannounced at Lescarbault's home. Without identifying himself, he launched into a brusque interrogation of the shy physician, demanding to know how Lescarbault came to the absurd conclusion that he'd seen "an intra-Mercurial planet." Lescarbault recounted the story, detail by detail. Yes, he'd had the presence of mind to make some hasty measurements of the speed and direction of motion of the mystery spot. Using an old clock and a pendulum with which he took his patients' pulse, he estimated the duration of the transit at 1 hour, 17 minutes, 9 seconds. Leverrier, now convinced that Lescarbault had indeed witnessed the transit of a previously unknown planet, revealed who he was and congratulated the bewildered man.

On his return to Paris, Leverrier used what the physician had told him to build up a picture of this alleged new world. It was, his calculations suggested, only about one-seventh the size of Mercury and went around the sun in a nearly circular orbit at a distance of 13 million miles, just over a third that of Mercury, tilted at about 12 degrees with respect to the plane of Earth's orbit. This hypothetical closest member of the sun's family, he reckoned, took only 19 days, 17 hours, to make one solar circuit–its year–and was never more than 8 degrees from the sun in the sky. That would make it almost impossible to see except during a transit and, perhaps, a solar eclipse, when most of the sun's glare was blotted out. On January 2, 1860, he announced the discovery of the new planet to a meeting of the Academy of Sciences in Paris. He suggested that the planet be named Vulcan, after the Roman god of fire.

Quickly, the astronomical world was abuzz with rumor and discussion about Leverrier's new world. Some doubted its authenticity right from the start. An eminent French astronomer, Emannuel Liais, who was working for the Brazilian government in Rio de Janeiro in 1859, claimed to have been studying the surface of the sun with a telescope twice as powerful as Lescarbault's at the very time Lescarbault said he witnessed the mysterious transit. Liais, therefore, was "in a condition to deny, in the most positive manner, the passage of a planet over the sun at the time indicated."[18]

Such denials, however, did nothing to stem the tidal wave of reports that reached Leverrier about unexplained transits–some referring to observations in the past (usually undated), many from amateurs, most unreliable. Despite the paucity of good data, Leverrier continued to tinker with Vulcan's orbital parameters as each new reported sighting arrived on his desk. Frequently he'd announce dates of future Vulcan transits, and when these failed to materialize, he'd twiddle the parameters some more.

There were the odd moments when it seemed Vulcan might have been genuinely sighted. In London, shortly after eight o'clock on the morning of January 29, 1860, the amateur astronomer F. A. R. Russell and three other people witnessed what they took to be a transit of an intra-Mercurial planet. A couple of years later, on March 22, 1862, again around breakfast time, another amateur observer, a Mr. Lummis of Manchester, England, reported a transit and alerted a colleague, who also followed the event. Based on their accounts, two French astronomers, Francoza Radau and Benjamin Valz, independently calculated the object's orbital period, coming up with values of 19 days, 22 hours, and 17 days, 13 hours, respectively.

Leverrier died in 1877, convinced to the end that Vulcan was real. With the passing of its arch proponent, Vulcanology went into decline. The casebook remained open for a while. During the total solar eclipse of June 29, 1878, two experienced astronomers, James Watson, director of the Ann Arbor Observatory in Michigan, and Lewis Swift, an amateur from Rochester, New York, both claimed to have seen a Vulcan-type planet close to the sun. Watson, observing from aptly named Separation, Wyoming, placed the planet about 2.5 degrees southwest of the sun, while Swift, from a vantage point near Denver, Colorado, saw what he believed was an intra-Mercurial planet about 3 degrees southwest of the sun. Swift estimated its brightness to be the same as that of Theta Cancri, a fifth-magnitude star, which was also visible during totality, about 6 or 7 arc-minutes from the mystery object. These two men were excellent observers, with a long track record of success. Watson had already discovered more than twenty asteroids; Swift had the scalps of several comets to his name. Both described the color of their hypothetical intra-Mercurial planet as red. Watson reported that it had a definite disk–unlike stars, which appear in telescopes as mere points of light. What they saw has never been explained, nor has it ever been seen again. But it was almost certainly not Leverrier's long-sought-for inner world.

Sunspots, background stars, telescopic ghosts–take your pick–these were among the culprits, in the opinion of most astronomers, behind the various reported sightings of Vulcan over the years. By 1896 a new director of the Paris Observatory, Félix Tisserand, had concluded there was no object close to the sun that could account for Mercury's puzzling perihelion advance. One of the final theories that might explain the extra 43 arc-seconds per century by which Mercury's point of closest approach drifted around the sun had fallen. That left the almost unthinkable: if Mercury was behaving the way it should, then maybe there was something wrong with the rest of the universe, or, at least, our understanding of it. Could gravity itself be off-kilter?

No one had seriously questioned Newton's theory of gravity for a long time. But with the apparent demise of Vulcan and of other possible disturbers of Mercury's path, the finger of suspicion began slowly to turn on one of the pillars of classical physics. Two American astronomers were among the first to ask questions about the version of gravity that had stood for more than two centuries. One of these was Simon Newcomb, a brilliant Canadian-born mathematical astronomer who'd earned a towering reputation for his precise determination of the movements of the sun, the moon, and the planets and their satellites. His solar system data provided the basis of nautical and astronomical almanacs published in the United States and Britain until as recently as 1984. The other doubter of Newtonian theory was Asaph Hall, an eagle-eyed discoverer of the two moons of Mars.

Both Newcomb and Hall went back to an issue that had been raised in the early days of Newton's universal law: what if gravity didn't obey a simple inverse square rule? Hall suggested that instead of varying as $1/r^2$, the actual formula for gravitation might involve a factor of $1/r^{2.00000016}$. To most scientists, that slight but annoying deviation from whole number perfection was about as desirable as a wart on the nose of the Mona Lisa. Yet there was no getting away from the fact that Newtonian gravity theory couldn't explain the anomalous meanderings of Mercury.

One stubborn gap in understanding was bad enough. But there were a number of other puzzles that seemed to be thumbing their noses at Newton's way of looking at gravity, and they also were giving scientists cause for concern.

Equality for the Masses

Baron Roland von Eötvös, or Vásárosnaményi Báró Eötvös Loránd, to give him his full Hungarian title (Roland is the Germanized form of Loránd), was a physicist who spent the last thirty years of his life deeply concerned about mass. It seems such a simple idea on the face of it: *mass* is the amount of stuff–the quantity of matter–an object contains. Concepts in physics don't get any more self-evident or commonplace than this. The trouble is, there isn't just one kind of mass, as you might expect. There are two, and they come about in completely different ways. But here's the kicker: although these two forms of mass seem to be totally unrelated, they turn out to have the same value. It was an absolute mystery to physicists why this should be.

The mystery stems back to a couple of facts about the world that Galileo established through his experiments. The first is that all objects, light and heavy, fall at the same rate in the absence of air resistance. The second is that objects have an innate property, called inertia, that keeps them moving at a constant speed or at rest unless acted on by a force. Then along came Newton. To the question, *Why* don't heavier objects fall faster than lighter ones? he gave a beautifully simple, but deeply puzzling, answer, "Because heavier objects also have greater inertia."

Determined to test this, let's say, you go to Pisa armed with an iron cannonball and a wooden ball that's the same size but ten times lighter. You stagger to the topmost balcony of that famous listing bell tower of the cathedral, lean precariously over the edge, and drop the two balls at the same instant. About two-and-a-half seconds later, you're relieved to see and hear them thud to the ground simultaneously. Galileo was right! And the reason, Newton says, is this: while the cannonball is pulled down with a force that's ten times larger, it also resists being moved with an inertia that's ten times greater. *These two effects exactly cancel out.* The greater mass of the iron ball means that it would accelerate faster, and so hit the ground sooner, than the wooden ball, were it not for its stronger reluctance to change its state of motion. This greater inertia of the metal ball exactly offsets any tendency to fall more quickly because of gravity.

The point is that although Newton was perfectly well aware of this canceling out effect of gravitation and inertia, it doesn't follow from his theory. On the contrary, it's just stuck in as a separate empirical fact–a vital but baffling truth of nature. Right in the first paragraph of his *Principia*, Newton acknowledged that only by experiment could he verify that the mass of an object that crops up in his laws of motion is directly related to its weight: "It [mass] can also be known from a body's weight, for–by making very accurate experiments with pendulums–I have found it to be proportional to the weight."

The idea behind Newton's pendulum experiments is quite simple. Take two pendulums of the same length but with different weights on their ends. Start them swinging together from the same height at the same time. If the pendulums stay in sync, then the period–the time for one complete swing back and forth–must be independent of the weight, from which it follows that the weights (a measure of their gravitational masses) are proportional to their inertial masses. This is exactly what Newton found. He also showed, by using substances as diverse as gold, silver, lead, glass, salt, and wood, that the relationship holds whatever substance is used as the material of the pendulum bob. Then, going beyond his experiments, he looked for evidence in the heavens for this proportionality of weight and mass. And his gaze fell upon the moons of Jupiter. If these moons were attracted to the sun by a force that was disproportionate to their masses in relation to the attraction of Jupiter itself to the sun, then their orbits wouldn't be stable. Yet, he pointed out, round and round they went, year in, year out, their orbits unchanging.

So, here was a strange situation. On the one hand, Newton's universal law of gravitation said that gravity was a property created and felt by anything that had mass. On the other hand, Newton's laws of motion–in particular, the second law–said that the mass of an object determines how much the object accelerates when it's given a push. Why should the laws of gravity and motion hinge upon the same property? Gravity and motion, after all, don't seem much alike. Why does mass do double duty as a measure of the resistance of an object to a change in velocity (the object's inertia) and as the property involved in reacting and giving rise to a gravitational force?

Newton himself, in the *Principia*, pointed out that this coincidence doesn't apply to forces in general. Think about the force of magnetism. That force isn't proportional to the mass of the attracted body. Nor does magnetism treat all materials the same. If magnetism stood in for gravity, the world would be a very different place: soft iron objects would fall quickly, and aluminum and stone ones slowly or not at all; meanwhile, some materials, such as bismuth or graphite, would rise very slowly. Magnets would go up or down depending on their orientation.

Newton spoke about the proportionality of an object's weight–the force with which it's gravitationally drawn toward another body–and the object's mass. He didn't specifically mention two different kinds of mass. Physicists today, though, do exactly that. They talk about "gravitational mass" as the property of an object that both creates and responds to gravity. And they talk about "inertial mass," which is the property that decides how easy it is to make an object move in a different way when no other forces are acting on it. What Newton's experiments with pendulums and Galileo's earlier experiments with rolling balls and inclined planes showed is that these two kinds of mass are equivalent–that there's an *equivalence principle* in nature. Or, to turn the idea around, a consequence of this equivalence principle is that all objects fall toward the Earth at the same rate.

Newton's pendulum experiments proved an equivalence of gravitational to inertial mass of better than one part in a thousand. That didn't leave much room for any future discrepancy to be found, but it left some. Equipment more advanced than Newton's might show that the two kinds of masses weren't exactly the same. It was to test this possibility that Baron von Eötvös and his colleagues, using an instrument of almost unbelievable sensitivity, launched an investigation in the early part of the twentieth century.

A Question of Balance

Born in Pest (one of three former cities that now form Budapest), Hungary, in 1848, the year of the Hungarian revolution, Eötvös was the son of a well-known poet, writer, and liberal politician–a cabinet minister at the time, who figured prominently in his country's intellectual life. Eötvös first studied law but soon switched to physics and went abroad to study in Heidelberg and Königsberg. Shortly after receiving his doctorate, he took a post as university professor in Budapest and rose to become a central player in Hungarian science for almost half a century.

Eötvös tended to work in areas that weren't always fashionable but that he personally found rewarding and potentially useful. "The true natural scientist," he said, "finds pleasure in research itself and in those results which help to increase the prosperity of mankind." He won international recognition for his groundbreaking work on capillary action–the effect by which liquids climb up narrow tubes or gaps–and for introducing the concept of molecular surface tension. His later work, however, focused mainly on gravity.

Eötvös was eager to probe, in finer detail than ever before, how Earth's gravitational pull varies from one point on the surface to another. Newton had known that the effective weight of an object was slightly less at lower latitudes than at higher ones because of the planet's rotation; the faster spin nearer the equator reduces the net downward pull. He had been aware that if a pendulum clock, which depends on gravity, were accurate in England, it would run a bit slower after being moved to the equator. But there were all sorts of subtler variations in pull, depending on local topography and geology, that were too small for a swinging pendulum to reveal. To detect these, Eötvös turned to a device that had been specially designed to measure small forces or tiny differences between forces–the torsion balance.

Twist a wire or a thread and it will resist with a force proportional to the twisting force, or torque, that you apply. This is the principle on which the torsion balance works. Basically, it consists of a wire fixed at one end and arranged so that a force applied at the other, free end tends to wind it out of shape.

The amount of twisting gives a superaccurate measure of the acting force.

Two people are credited with inventing the torsion balance, completely independently of each other. One was the French physicist Charles de Coulomb, who, in 1777, built and used his device for investigating the electrical force between small charged spheres. The other was the English geologist, astronomer, and vicar John Michell, who is generally considered the father of seismology (the science of earthquakes) and whom we'll meet again later for a very surprising reason (see chapter 10). Quite when the idea of the torsion balance popped into Michell's mind isn't clear; it may have been as early as 1750 and certainly predated Coulomb's experiments. Eventually, Michell built an instrument similar to Coulomb's but designed to measure the attraction not between electrical charges but between *gravitational* "charges"–two pairs of masses in the form of metal balls suspended close together.

Michell's goal for his instrument was nothing less than to weigh the Earth. That may seem farfetched–one has visions of putting the planet on a giant set of scales. But weighing the Earth really boils down to finding an accurate value for G, the *universal gravitational constant,* which appears in Newton's formula for the force between two masses. If the values of these two masses are known, together with the distance and the gravitational force between them, then G can be calculated straight from the equation. With the value of G known, it can be plugged back into Newton's formula as applied to the force between Earth and an object of known mass, in other words, the object's weight, to give the mass of Earth. The trick to weighing Earth, then, is to be able to measure accurately the force of gravity between two small masses–hence, the torsion balance.

Unfortunately, Michell died in 1793 before he had a chance to put his Earth balance into action. The pieces of his apparatus passed safely into the keeping of Cambridge physicist Francis Wollaston, who, being otherwise occupied, handed them on to an old friend of his and Michell's, Henry Cavendish, the grandson of the second duke of Devonshire.

Cavendish was affluent, quiet, solitary, and, frankly, a bit odd. He attended Cambridge University from 1749 to 1753 but left without taking a degree. Thanks to a generous inheritance, he was able to pursue his fascination for science, unhindered by having to work for a living, in a laboratory he set up in his own home. Doubtless this seclusion suited him; by one account, he was so shy, especially of women, that he had a back staircase added to his house in order to avoid running into his housekeeper. When he died, he left a large estate to his old university, which was used to endow the now-famous Cavendish Laboratory in 1871.

In his home lab, Cavendish rebuilt most of Michell's instrument and added some refinements of his own. The finished product consisted of a six-foot wooden rod, suspended by a metal fiber, with two-inch-diameter lead balls mounted on the rod at either end. At the start of the experiment, he brought up two 350-pound lead balls, one beside each of the smaller ones, so that the massive balls would tend to attract the smaller ones and twist the rod clockwise. After leaving the whole thing for some hours to settle down, he brought the massive lead balls to the other side of the small ones in order to twist the rod counterclockwise. By measuring the angle between the two different positions of the rod, Cavendish was able to work out the attractive force exerted by each of the large lead balls on the smaller ones. The whole apparatus was enclosed in a draft-proof room to guard against the influence of air currents, and the readings were taken through telescopes mounted on each side.

Cavendish carried out his famous experiment in 1797 at the age of sixty-seven, and published his results the following year. But he wasn't the first person to weigh the Earth. That honor had already gone to Astronomer Royal Neville Maskelyne, who in 1774 used a method called the Attraction of Mountains to come up with a figure of about 5 million trillion metric tons. Cavendish's far more precise work upped that value to almost 6 million trillion metric tons, corresponding to an average density for Earth of 5.48 times that of water–remarkably close to the modern value of 5.52. With Earth's mass accurately pinned down, the sun's mass was easy to calculate from the size and period of Earth's orbit.

Cavendish's extraordinary attention to detail and his careful bookkeeping of all the errors in his method led some scientists to describe this experiment as the first modern one in physics. It was certainly the first to provide experimental, as distinct from purely observational, support for Newton's law of gravity. And it was one of the inspirations for Baron von Eötvös as he sought to map the delicate spatial fluctuations in our planet's gravitational pull.

Eötvös started to experiment with gravity and the torsion balance around 1885. His first instruments were similar to those of Cavendish and Coulomb, with a horizontal rod suspended at the center, but of greater sensitivity, and the main focus of his work was geophysical. By 1890, he'd been able to measure the mass of the Gellért-hegy–a hill above Budapest that rises steeply, almost clifflike, from the Danube. But he was intrigued, too, by the more fundamental issue of the proportionality of inertial and gravitational masses, and, also in 1890, he reported his initial findings on this in the *Proceedings of the Hungarian Academy*.

Quite why Eötvös began a set of experiments to test the equivalence principle isn't clear. He certainly recognized both the importance of this principle for Newtonian mechanics and the fact that previous experimental evidence was very limited. But it was probably only when he grasped the astonishing capabilities of his instrument that he decided to go ahead. Eötvös realized that he could test the equivalence principle by placing objects of different materials at opposite ends of his balance. The net force acting on each object is a combination of Earth's gravitational attraction, which is proportional to the gravitational mass of the object, and the centrifugal force due to the planet's rotation, which is proportional to the object's inertial mass. If two different materials were put on the balance and the ratio of the gravitational mass to the inertial mass for one didn't equal that ratio for the other, the balance would rotate. No rotation was observed, however. In his 1890 report, Eötvös announced that he'd improved on the precision of the most careful pendulum experiments ever carried out–by Wilhelm Bessel in the late 1820s–by a factor of four hundred. Yet, better was to come.

The following year, he crafted a new kind of torsion balance, now known simply as the Eötvös balance, in which only one weight hangs down from the end of the rod. Also called a horizontal variometer, because it makes it possible to measure how quickly the acceleration due to gravity changes between neighboring points on Earth's surface, it was widely used for charting local variations in gravity and later for prospecting for oil and natural gas. For more than a decade, Eötvös did no more work on the equivalence problem. Then, in 1905, he and his assistants Jenõ Fekete and Dezsõ Pekár began a far more extensive series of investigations that involved taking data with the new type of balance for about four thousand hours over a three-year period. At the Sixteenth International Geodesic Conference in London in 1909, Eötvös announced that his team had now established that gravitational mass and inertial mass were equivalent to within one part in 200 million. At less than this limit, he found discrepancies between different types of material, but he put these down to experimental error.

The equivalence principle had been proven beyond all but the tiniest sliver of doubt. But *why* was it true? It's easy to imagine a universe in which gravitational mass and inertial mass aren't the same–a universe in which, for example, a chunk of lead has (as in our own universe) about four times the weight of a chunk of aluminum of the same size and shape but only twice the inertia. Newton's theory had no answer to the riddle of equivalence; it could only shake its head and accept it as an extraordinary fact.

The Unexplained Force

Another mystery that hung over Newton's theory–a profound, unsettling mystery–concerned gravity's very nature. Gravity acts at a distance, even across the void of space, between all objects that have mass. But the great puzzle is: how do those objects know that the others are there? Put another way, how is the effect of gravitation communicated? For Galileo, for Newton, and for all other scientists up to the beginning of the twentieth century, *gravity* was no more than an empty name for a phenomenon they didn't really understand. In a letter to the Cambridge theologian Richard Bentley, Newton wrote, "You sometimes speak of gravity as essential and inherent to matter. Pray do not ascribe that notion to me, for the cause of gravity is what I do not pretend to know and would therefore take more time to consider of it."

Newton had shown that gravity works throughout the universe as it does here on Earth. He had described it mathematically. But he hadn't, he fully admitted, grasped its essence, its origin, its modus operandi: "It is inconceivable that inanimate brute matter should . . . affect other matter without mutual contact. That gravity should be innate, inherent, and essential to matter, so that one body may act upon another at a distance, is to me an absurdity that I believe no man who has in philosophical matters a competent faculty of thinking can ever fall into it."

By the second half of the nineteenth century, physicists spoke of two great forces in nature. One was gravity, the other electromagnetism–a marriage of electricity and magnetism. Both gravity and electromagnetism were thought to obey the inverse square law and to act across distance with a range that was essentially unlimited. Yet, although these forces seemed to have a great deal in common, their status in physics was very different. Scientists believed they knew how electromagnetism worked. The great Scottish physicist James Clerk Maxwell had boiled down the behavior of all electric and magnetic phenomena to nine fundamental equations, later reduced to just four. From these equations he'd deduced a remarkable fact, namely, that any movement of a magnet or an electric charge, such as a current flowing through a wire, will give rise to waves that travel out with a speed of 300,000 kilometers per second. Since this value was also known to be the speed of light, it quickly became clear that light must be a form of electromagnetic wave and, furthermore, that there must be other forms of electromagnetic wave with both higher and lower frequencies than those of visible light.

Now, if you're a Victorian scientist, you have definite views about waves. One thing you believe is that a wave can travel only if it's in a medium it can make vibrate. In the case of an ocean wave, what vibrates is the water or, more precisely, the surface of the water. For sound waves, the vibrating medium is the air. But there seems to be a problem when it comes to light because light–for example, from the sun–can travel through space. To Maxwell and his contemporaries, there was only one solution to this difficulty. What appears to be empty space isn't really empty at all but is filled with a mysterious, intangible substance–the ether. Maxwell wrote an article on the ether for the 1878 edition of *Encyclopaedia Britannica* in which he discussed some of its wonderful properties. It must be elastic enough to support the vibrations of electromagnetic waves, yet be virtually weightless and offer no resistance to bodies passing through it. The ether was championed as the means by which not only light but also heat, electricity, and magnetism were able to travel and cause effects across the otherwise pure vacuum of space. It was natural to ask whether the ether might explain, too, how bodies in the solar system and beyond could interact gravitationally.

In *A Theory of the Electromagnetic Field* (1864), Maxwell wondered if his theory could be modified to describe gravity: "After tracing to the action of the surrounding medium both the magnetic and the electric attractions and repulsions, and finding them to depend on the inverse square of the distance, we are naturally led to inquire whether the attraction of gravitation, which follows the same law of the distance, is not also traceable to the action of a surrounding medium."

But here he ran into a problem–a paradox caused by the fact that with gravity there's only attraction, never repulsion. In gravity's case all "charges" are like and all try to pull each other together. This made Maxwell's equations go awry. The upshot of the attraction of like bodies is that the energy of the surrounding medium–the ether or whatever–is decreased by the presence of other bodies. "As I am unable to understand in what way a medium can possess such properties," Maxwell concluded, "I cannot go further in this direction in searching for the cause of gravitation."

Newton's theory of gravitation had been wonderfully successful. But as the nineteenth century drew to a close, it faced challenges on multiple fronts. With the failure to find Vulcan, there was the unsolved problem of Mercury's anomalous orbit. There was the curiously ad hoc nature of the equivalence principle. And there was the disturbing absence of any kind of model or mechanism to explain *how* gravity actually worked. Newton had transformed our view of the universe. Now a new revolution was fomenting that would change forever our notions of mass and energy, space and time.

7

When Gravity Became Geometry

Newton, forgive me.

—ALBERT EINSTEIN

A total eclipse of the sun is a magical event a few minutes of eerie gloom when the dazzling disk of the sun gives way to pale tentacles around a heart of darkness. For many, the spectacle is a once-in-a-lifetime experience. For a handful of scientists in 1919, however, it was more than that. It was an opportunity to change the world, or our perceptions of it, forever.

Two expeditions, a redundancy designed to limit the risk of bad weather, had set out that year from England to points along the predicted path of the total eclipse, one led by Arthur Eddington to the island of Principe off the West African coast, the other, under Andrew Crommelin of the Royal Observatory at Greenwich, to Sobral in Brazil. Eddington was the mastermind behind the venture. His goal was put to the test a new way of thinking about gravity.

The fateful day of May 29 dawned with torrential rain at the African site. Meanwhile, Crommelin in Brazil had problems of his own, having focused his main telescope the night before only to find that a rise in temperature the next day gave the instrument a blurred view by the time of the eclipse. Fortunately, his backup scope performed well. Across the Atlantic, the weather in Principe partly cleared by the afternoon of the great event, enabling Eddington, too, to capture some useful pictures of the eclipse. What the results showed made front-page news around the world. The faint images of stars seen close to the sun during the total eclipse were ever so slightly out of position–displaced by an amount that would have baffled Newton but that completely vindicated the predictions of one Albert Einstein. Overnight, a new genius of science had become a household name.

Ether Or . . .

What would it be like to travel at the speed of light? With that simple question an anonymous worker in a patent office in Switzerland opened the door to a deeper understanding of the universe and, in time, to a mind-bending new theory of gravity.

Einstein is forever pictured as the gentle-faced, white-haired inventor of relativity theory. But two aspects of that image are misleading: Einstein did most of his important work before he was thirty-six–by middle age the flame of creativity had deserted him–and he was by no means the sole contributor to what became known as the special theory of relativity.

Three hundred years earlier, Galileo had been the first to throw out the old absolutist views of Aristotle. Motion, or at least steady motion in a straight line, Galileo showed, had meaning only in relation to something else. He imagined being in the windowless cabin of a sailing ship with only some birds and a few fish in a bowl for company. If the ship were in calm waters it would be impossible to tell, said Galileo, whether the ship was at rest or, in the absence of any swaying or pitching, moving along with an even speed. This "principle of relativity" is enshrined in Newton's formulation of physics. But it's important to understand exactly what Newton believed on this issue. He was firmly convinced that objects can have absolute velocities–that some things *really* are at rest while others really are in motion. However, he was also adamant that there was no way to measure these absolutes. The best an observation can do is tell you the velocity of something relative to your own velocity, or a position relative to your own position. All the laws of mechanics, he argued, work identically, no matter how you're moving. It's a claim that seems to stand up well in everyday situations that involve solid objects.

But then along came the discovery of electromagnetic waves by James Maxwell, and suddenly a spanner was thrown into the works. If these waves needed an all-pervasive medium–the ether–to allow their passage, then it seemed that this medium would let back in the idea of a detectable absolute frame of reference–one that's stationary with respect to the ether.

If the ether existed, it ought to be possible to measure its effects, and Maxwell had an idea how. In 1879 he wrote to David Todd, the director of the Nautical Almanac Office in Washington, D.C., asking whether the eclipses of Jupiter's moons could be used to detect Earth's motion through the ether. His idea was based on a method for determining the speed of light that had been used by the Danish astronomer Ole Römer a couple of centuries earlier. Römer found that the intervals between eclipses of the inner Jovian moon Io occurred twenty-two minutes later–the modern figure is about sixteen minutes–when Earth was furthest away from Jupiter in its orbit than six months later when it was closest. This time difference he rightly attributed to the fact that it takes light longer to reach us when it has to cross the diameter of Earth's orbit. Maxwell argued that if Earth is moving through the ether, then the speed of light, presumably, would be less when the light was swimming upstream, against the flow of the ether, than if it were moving with the ether's flow. He calculated that Römer's time delay should vary by up to a second as Jupiter orbited the sun. Todd wrote back with bad news: astronomical data available at the time, unfortunately, wasn't accurate enough to make the experiment viable.

The story, though, doesn't end there. A young American at the U.S. Naval Academy, Albert Michelson, came across the correspondence between Maxwell and Todd while on study leave in Germany. A few years earlier, at the age of just twenty-five, Michelson had obtained a value for the speed of light–299,910 kilometers per second–within a whisker of today's accepted best value. After turning the problem over, he thought he saw a way to detect Earth's motion through the ether. With financial backing from Alexander Graham Bell, he built an instrument with mirrors, a small telescope, and a light source to compare the speed of light in two directions at right angles. If there were a wind caused by Earth's motion through the ether, then it ought to show up as a discrepancy between the light-speed in these different directions. But, to his surprise, Michelson found no hint of such a discrepancy. In 1881 he reported, "The result of the hypothesis of a stationary ether is shown to be incorrect, and the necessary conclusion follows that the hypothesis is erroneous." Urged on by one of the great elders of science, Lord Kelvin (William Thomson), Michelson and a colleague, Edward Morley, devised a much more sensitive version of the earlier experiment–but again they drew a blank.

Physicists were now becoming deeply puzzled about what was going on. If the ether existed, why didn't motion through it affect the speed of light? After all, to make a familiar comparison, if someone rows a boat with or against the flow of a river, he travels at a different rate than if he rows directly across the river from bank to bank. In 1889, the Irish physicist George Fitzgerald wrote a short paper[13] in which he made a seemingly bizarre claim: "[T]he length of material bodies changes, according as they are moving through the ether or across it, by an amount depending on the square of the ratio of their velocities to that of light." Was Fitzgerald mad? Objects changing length depending on the direction they're moving? This sounded more like wizardry or desperation–than science. Yet it was the only way, at least on paper, to reconcile the null results of the Michelson-Morley experiment with the supposed presence of the ether.

The Irishman wasn't alone in sticking his neck out on this issue. Three years later the Dutch physicist Hendrik Lorentz, completely unaware of Fitzgerald's paper, proposed an almost identical length contraction in a paper of his own. When it was pointed out to Lorentz in 1894 that Fitzgerald had published a similar theory, he wrote to Fitzgerald, who replied that he was glad to know that Lorentz agreed with him, "for I have been rather laughed at for my view over here."

As well as a length contraction, Lorentz suggested there would be another outrageous effect for any objects or observers along their direction of motion: time would pass more slowly. The faster someone or something moved, the shorter they would get and the slower their clocks would tick, relative to an observer who was standing still with respect to the ether. The complete set of equations that describe these weird changes has come to be known as the Lorentz transformation, even though Lorentz wasn't actually the first person to write the equations down. That honor went to the German physicist Woldemar Voigt in 1887 while was doing work on the Doppler shift–the change in wavelength caused when a source is moving away from or toward an observer. Voigt corresponded with Lorentz about the Michelson-Morley experiment in 1887 and 1888, but Lorentz doesn't seem to have learned of the transformation at that stage.

By the turn of the century, physicists were beginning to realize that our conceptions of space and time were in need of a serious makeover. Lorentz had grasped that the transformation named after him was central to a full understanding of Maxwell's equations. Only by applying this transformation to Maxwell's equations, he realized, could they be made to square with experimental results in the case of moving bodies. Meanwhile, Henri Poincaré, the great French mathematician and physicist, was asking questions about the comparison of time intervals and the meaning of simultaneity. In his paper "La mesure du temps" (The measurement of time), which appeared in 1898, Poincaré says, "[W]e have no direct intuition about the equality of two time intervals. The simultaneity of two events or the order of their succession, as well as the equality of two time intervals, must be defined in such a way that the statements of the natural laws be as simple as possible."[30]

By 1900 doubts were also being openly expressed about the existence of the ether. In 1904 Poincaré concluded a lecture with these prophetic words: "Perhaps we must construct a new mechanics of which we can only catch a glimpse . . . in which the velocity of light becomes an unpassable limit." The scene was set for the arrival on the world's scientific stage of an unlikely genius.

Transformations

Einstein didn't learn to talk until he was three and never rose above mediocrity in his exams at school and college. But this slow start in life, he pointed out, brought its eventual rewards: "The normal adult never bothers his head about space-time problems. . . . I, on the contrary, developed so slowly that I only began to wonder about space and time when I was already grown up. In consequence I probed deeper into the problem than an ordinary child would have done."

As a sixteen-year-old, he started pondering the kind of thought experiments that would become a hallmark of his career. What would you see, he wondered, if you rode astride a light beam or held a mirror and looked into it while traveling at the speed of light? Over the next ten years, he continued to question the nature of space, time, and light and to ask how physics would need to change if it were to address all the circumstances in which objects are moving relative to one another.

Amazingly, he brought about a revolution in physics while working as a technical expert, third class, at the Bern patent office. Isolated from academe, unheard of by eminent professors at the venerable universities of Europe, he whiled away his spare hours rethinking the notion of time and mulling over the relationship between the speed of light and the speed of an observer. During this period of anonymity, his closest confidante was Michele Besso, a fellow patent clerk and keen violinist, whom Einstein had first met at a musical evening in 1896. Of Besso, Einstein said, "I could not have got a better sounding board in the whole of Europe."

For over a year, Einstein had been stymied by a problem he couldn't see a way around. It was simply this: if the speed of light is a universal constant, as he guessed it was, then the usual "Galilean" addition of velocities doesn't work. According to the physics of Galileo and Newton, if you walk down the aisle of a train that's moving along the track, then your speed relative to the ground equals the train's speed plus your walking speed. This seems like perfectly good common sense, and it is what you'd actually find if you took measurements, unless you could measure with atomic precision, in which case, you'd notice a puzzling discrepancy. In the same way, if the train has a lamp at the front, then, according to Newtonian science, the speed of the rays leaving the lamp ought to equal the train's speed plus the speed of light. But if the speed of light doesn't depend on how the source is moving, as Einstein believed, this simple addition can't be right; otherwise, you could easily break the light barrier.

Then, "one momentous day in May," recalled Einstein, inspiration dawned. He visited Besso and explained what was on his mind. "Today I come here," he said, "to battle against that problem with you." Together they looked at the question from every angle until, suddenly, Einstein saw the answer to his difficulty: if the speed of light is always the same, then space and time must be changeable. The notion that space and time formed a fixed backdrop against which the drama of the universe played out–Newton's viewpoint–was wrong. Space and time, length and duration, would be measured differently depending on the relative motion of the observer.

On June 30, 1905, Einstein's paper that formed the basis of what would eventually be called the special theory of relativity was published. Modestly titled (translated from German) "On the Electrodynamics of Moving Bodies," it was very unusual as scientific papers go.[8] No references appeared in it to any other papers or to the work of any other researcher; there is merely an acknowledgment of the "loyal assistance of my friend M. Besso." Of it, the English author and physicist C. P. Snow has said, "There is a good deal of verbal commentary. The conclusions, the bizarre conclusions, emerge as though with the greatest of ease: the reasoning is unbreakable. It looks as though he had reached the conclusions by pure thought, unaided, without listening to the opinions of others. To a surprisingly large extent, that is precisely what he had done."[35]

There's a directness about the June 1905 paper, a clarity, that comes across without the need for advanced math. The seemingly ad hoc transformations of Lorentz and Fitzgerald follow naturally and inevitably, in Einstein's hands, from simple geometry and the well-known theorem of Pythagoras. And all of this simplicity and clarity–in fact, the whole theory of special relativity–stems from just two basic postulates that Einstein lays down early on. The first is that the laws of physics are identical in all inertial frames; that is, to all observers who are at rest or traveling with constant velocity relative to one another. The second is that the speed of light is the same in any inertial frame, whether the light is given off by a body at rest or in a state of uniform motion.

Right at the start of his paper, Einstein does some impressive spring cleaning–dispensing with the ether along with any lingering notions of absolute space and an absolute state of rest: "[T]he introduction of a light ether will prove to be superfluous since, according to the view to be developed here, neither will a space in absolute rest endowed with special properties be introduced nor will a velocity vector be associated with a point of empty space in which electromagnetic processes take place."

Einstein restored the principle of relativity to physics because he insisted that physical laws have to look the same to all observers traveling at constant velocity–exactly as Galileo and Newton had done. But the only way he could do this *and* keep the speed of light constant in his theory, no matter what the motion of the source, was to allow space and time to be flexible. From Einstein's postulates, the Lorentz transformation flows as a matter of course; time and length really do shrink along the direction of movement, although these "relativistic" effects only kick in at velocities that are a significant fraction of that of light itself.

Oddly enough, Einstein never talked about the Michelson-Morley experiment, even though his seminal paper gave a complete explanation of its null results. Despite being the main reason that other scientists, such as Poincaré and Lorentz, were questioning the established notions about space and time, the experiment was never acknowledged by Einstein as having any influence on his thinking. Even more strangely, Poincaré and Einstein seemed to go out of their way to avoid crediting each other.

Poincaré was closing in on relativity theory himself at the time Einstein achieved his breakthrough and perhaps, with the help of others, would have come to similar conclusions sooner rather than later. Yet Einstein mentioned him only once in all of his papers, while Poincaré never wrote a word on relativity in which he referred to Einstein. Lorentz, on the other hand, was praised by both and often cited by them, even though Lorentz never fully accepted relativity. Here is the man after whom the transformation equations at the heart of special relativity are named speaking in 1913:

> As far as this lecturer is concerned, he finds a certain satisfaction in the older interpretation according to which the ether possesses at least some substantiality, space and time can be sharply separated, and simultaneity without further specification can be spoken of. Finally, it should be noted that the daring assertion that one can never observe velocities larger than the velocity of light contains a hypothetical restriction of what is accessible to us, a restriction which cannot be accepted without some reservation.

Long before he uttered these words, however, Lorentz had found himself in a minority. Acceptance of the special theory of relativity was increasing, especially following Max Planck's endorsement of it.

On September 27, 1905, Einstein sent out a sequel to his June paper in which he proved what has become the most famous formula in all of science.[9] Previously, it had been thought that energy couldn't be created or destroyed, a claim known as the law of conservation of energy. A similar conservation law was believed to hold true for mass. But what Einstein showed, again starting from his two basic postulates, is that energy and mass are interrelated–the one can turn into the other. How much energy you get for a given investment of mass is decided by the equation $E = mc^2$, where c is the speed of light. Because c^2 (c times c) is, when measured in conventional units, a very large number, a tiny amount of mass yields a vast outpouring of energy. In time, this relationship would prove to be the secret behind the energy production of stars and the key to the most destructive weapons ever conceived by humankind. It would play a part, too, in Einstein's formulation of a new theory of gravity.

Einstein melded energy and mass with his $E = mc^2$. But it was one of his old university professors, the Lithuanian-born mathematician Hermann Minkowski, who showed that space and time were also inextricably connected. When Einstein was an undergraduate at the Swiss Federal Institute of Technology in Zurich (also known as the Eidgenössische Technische Hochschule, or ETH) in the late 1890s, Minkowski was less than impressed by him. "A lazy dog," he called him, because of his apathy toward work. So no one was more surprised than Minkowski when his miscreant of a student made good: "Oh, that Einstein, always cutting lectures–I really would not have thought him capable of it." But it didn't take Minkowski long to grasp the importance of relativity theory and to plunge into it with wholehearted enthusiasm.

According to the Lorentz transformation, length contracts as time slows down or "dilates," suggesting a complementary relationship between the two. Minkowski took this idea to its natural conclusion: time (t) is just another coordinate like the three coordinates of space (x, y, z), so that every happening or "event" can be uniquely pinpointed in space and time by a quartet of values–its coordinates (x, y, z, t) in the continuum of space-time. Minkowski summed up these ideas in a lecture he gave in September 1908: "Henceforth, space by itself and time by itself are doomed to fade away into mere shadows, and only a kind of union of the two will preserve an independent reality."

In this strange but elegant new world of Minkowski space-time, we're each travelers in a four-dimensional realm. At each moment you exist at a specific point in space-time, and your life, from birth to death, can be charted as a unique trajectory through space-time–a complicated twisting path that Minkowski called a world-line. If you move relative to others, space and time change for you as determined by the Lorentz transformation. Move very quickly indeed, at speeds that are a sizable fraction of the speed of light, and these changes become marked: space is effectively traded for time–length contracting as time dilates, while energy of motion turns into increased mass. The mathematician sees these effects in terms of symmetry operations: the Lorentz transformation is equivalent to rotations and translations in the multidimensionality of space-time.

A Fall through Space-time

By the early years of the twentieth century, physicists had begun to appreciate, through relativity theory, the intimate connections between energy and mass and space and time. Yet gravity stubbornly remained outside this picture. Poincaré, in a paper submitted in July 1905, just days before Einstein's special relativity paper, suggested that all forces ought to transform according to the Lorentz transformation. But if this were the case, he pointed out, then Newton's law of gravitation couldn't be valid because Newton's law allows *instantaneous* action at a distance. Drawing an analogy with electromagnetic theory, Poincaré proposed that gravitational interactions take place at the speed of light and involve waves that propagate at this fixed rate. Lorentz, in 1900, had also hinted that gravitation could be put down to actions that travel at light-speed.

In 1907 Einstein began seriously to look into the problem of gravity. Two years after putting forward the special theory of relativity, he was sitting in his patent office in Bern wondering what would have to be done to Newtonian gravitation to make it fit in with his newly hatched theory. Suddenly, he recalled, he had "the happiest thought" of his life: "[F]or an observer falling freely from the roof of a house there exists–at least in his immediate surroundings–no gravitational field. Indeed if the observer drops some bodies then these remain to him in a state of rest or uniform motion. . . . The observer therefore has the right to interpret his state as 'at rest' [at least until he hits the ground!]."

To drive home this point, imagine a slightly different situation. You're in a windowless room and are told that one of two circumstances is true: either the room is floating in space far away from any source of gravity or it's an elevator whose cable has been cut. Your task is to decide which, without leaving the room or otherwise obtaining information from outside. According to Einstein, the task is impossible because there is no experiment you can carry out that will help you decide between the two scenarios. Nor, for the same reason, could you tell whether you were in a room that was sitting on Earth or being smoothly accelerated at 32 feet per second per second–the rate at which things fall freely in Earth's gravity–by a rocket.

There is simply no observable difference, Einstein realized, between acceleration and gravity. On some deep level, they're one and the same. Consequently, he said, "[W]e shall . . . assume the complete physical equivalence of a gravitational field and the corresponding acceleration of the reference frame. This assumption extends the principle of relativity to the case of uniformly accelerated motion of the reference frame."

This is an assumption that broadens the equivalence principle we met in the last chapter. That principle, remember, asserts, as Galileo showed, that all objects fall at the same rate, with the result that mass measured gravitationally is indistinguishable from mass measured by its inertia. What's now called the Einstein or strong equivalence principle goes beyond this older, weaker version by stating that *all* the laws of physics, not just the law of gravity, are the same in all small regions of space, regardless of their relative motion or acceleration.

In the same year, 1907, that Einstein announced this broader principle of equivalence, he also began linking his equation $E = mc^2$ with gravity. It had long been known that gravity acts on everything with mass. Now that mass and energy turned out to be two sides of the same coin, it seemed reasonable to Einstein that gravity could act on energy too. In particular, it ought to be able to influence the movement of light rays.

Einstein's very first scientific paper, published in March 1905, had been on the nature of light. In it, he argued that a well-known phenomenon in physics called the photoelectric effect could be explained if light behaved as if it consisted of tiny discrete particles. Later, these particles came to be known as photons. Because photons contained energy, and therefore, from the $E = mc^2$ relation, an equivalent mass, their paths ought to be bent by gravity, just as the path of a bullet is curved by gravity as it travels from gun to target. But in 1907, when Einstein realized this, he was thinking only in terms of how light might be influenced by gravity here on Earth, and there seemed little chance of experimentally verifying an effect that would be so small.

For four years, Einstein published nothing else on gravity. Then, in 1911, it dawned on him that the bending of light by gravity could be checked by astronomical means. Light from a background star ought to follow not a straight line but a gentle arc as it passed close to the sun as seen from Earth. Einstein came up with a figure for this bending, completely unaware that the same answer had been obtained back in 1803 by a little-known Bavarian astronomer named Johann van Soldner, who used Newtonian gravitational theory and treated light as a stream of little projectiles. In 1913, Einstein wrote to the American astronomer George Hale to ask if it were possible to look for the minuscule deflection of starlight by the sun without waiting for a total eclipse. Hale replied that it wasn't; the sun's blindingly bright disk needed to be completely blotted out before any deflection of starlight would show as an apparent displacement of stars from their normal positions. The German astronomer Erwin Finley-Freundlich planned an expedition to Russia to observe an eclipse due to occur there in 1914 and thus to test Einstein's prediction. But World War I intervened, and the expedition was canceled. For Einstein it proved to be a lucky break–his prediction would have turned out wrong.

By 1912 Einstein was hot on the heels of a new theory of gravity that would incorporate his strong equivalence principle. By calling on this principle, he realized, he could avoid dealing with gravity as a force altogether. Move in the right way, by free-falling, and you don't feel gravity: in an inertial frame, you're weightless and gravity drops out of the picture. But Einstein also realized that the Lorentz transformation of special relativity wouldn't carry over to a more general setting because the way you have to move to cancel out gravity is different in different locations. What he needed was some mathematical way to stitch together local inertial frames in different places so that gravity canceled out everywhere. Although he wasn't yet sure what form his new theory of gravity would take, he did know this: "If all accelerated systems are equivalent [with respect to the laws of physics], then Euclidean geometry cannot hold in all of them."

Euclidean geometry is the geometry we learn in high school, with its familiar straight lines, circles, and triangles. It's the geometry of the plane, or "flat space," and was fully described around 300 B.C. by Euclid in his monumental book *Elements.* Euclid started out by listing five axioms, or self-evident truths, together with five postulates, or additional assumptions. The last of these postulates, which has come to be called the parallel postulate, has always been a bit of an oddball. One way to state it is that given any straight line and any point not on it, we can draw through that point one, and only one, straight line parallel to the given line. On the face of it, this seems commonsensical and obvious (try it with pencil and paper). But there had always been a lingering doubt about whether the properties of parallel lines as presupposed in Euclidean geometry could be derived from the other postulates and axioms or whether the parallel postulate had to be assumed as an extra fact. In the early 1800s, three mathematicians, working independently, found good reason for this doubt. Remarkably, they discovered geometric systems that satisfied all the axioms and postulates of Euclidean geometry *except the parallel postulate.* These geometries showed not only that the parallel postulate must be assumed in order to obtain Euclidean geometry but, more importantly, that other geometries–non-Euclidean geometries–can and do exist.

Dreamworlds

The first to hint that there were geometric realms undreamed of by Euclid was the mighty Karl Gauss, German mathematician, astronomer, and physicist, who in 1817 wrote: "I became more and more convinced that the necessity of our [Euclidean] geometry cannot be demonstrated. . . . [W]e must consider geometry as of equal rank, not with arithmetic, which is purely logical, but with mechanics, which is empirical." In other words, argued Gauss, the geometry of the space we live in can't simply be assumed to be Euclidean; its nature must be determined by measurement and experiment. And this is exactly what he did. Commissioned by the government in 1827 to make a survey map of the region for miles around Göttingen, Gauss found that the sum of the angles in his largest survey triangle was different from the expected, Euclidean 180 degrees. The observed deviation–almost 15 arc-seconds–was both inescapable evidence for, and a measure of, the curvature of the surface of Earth. It was also the first concrete proof of a world that lay beyond Euclid's ken.

Gauss had many brilliant ideas that he didn't publish, and his pioneering thoughts on non-Euclidean geometry were among them. His motto, *pauca sed matura* (few but ripe) and his fear of "the clamor of the Boetians"–a reference to the people from a region of ancient Greece famous for their obtuseness–conspired to keep him silent on this topic. Only many years later, after his death in 1855, did the diary come to light in which Gauss had written down his manifesto for a non-Euclidean revolution.

The first mathematician actually to go to press with his

views on the subject was the Russian Nikolai Lobachevsky in 1826. He described a geometry in which Euclid's parallel postulate isn't obeyed and in which the sum of the angles of a triangle adds up to *less than* 180 degrees. This kind of geometry is said to be hyperbolic–the sort found on the surface of a saddle. Unbeknownst to him, a young Hungarian mathematician, János Bólyai, had made the same startling breakthrough a few years earlier. Bólyai could hardly believe what he'd found: "Out of nothing I have created a strange new universe." His father, Wolfgang, a friend of Gauss, had spent much of his life trying to prove Euclid's fifth postulate and reacted with alarm to János's revelation: "For God's sake, I beseech you, give it up. Fear it no less than sensual passions because it, too, may take all your time, and deprive you of your health, peace of mind, and happiness in life."

Gauss, however, reassured the elder Bólyai that the concept of geometries beyond that of Euclid wasn't as insane as it sounded and that, in fact, he'd held similar beliefs for several years. Finally and reluctantly, Wolfgang included his son's revolutionary work on geometry as an appendix to a book he published in 1832.

None of these contributions to exploring the non-Euclidean landscape had much effect on mathematics in the first half of the nineteenth century; the ideas were too arcane and bizarre, too heretical. Yet their time was coming. In 1853, when Gauss was seventy-six, his star pupil, Bernhard Riemann, had to give a lecture at the University of Göttingen to confirm his position as a faculty member. It was the tradition in such circumstances to offer three possible topics, but the choice of topic would be made between only the first two. Not surprisingly, given this normal course of events, Riemann hadn't fully prepared for his third choice: the foundations of geometry. Gauss, however, couldn't resist the prospect of hearing his wunderkind speak on a subject that he (Gauss) had grappled with for much of his life, and so he asked Riemann to deliver his third topic. After several postponements, Riemann gave his lecture "On the Hypotheses Which Lie at the Foundation of Geometry" in June 1854. It proved to be a triumph and marked a turning point in our understanding of non-Euclidean math.

Earlier in his career, Gauss had published results in which he hugely advanced the theory of surfaces in two dimensions. He'd shown that it isn't necessary to consider a two-dimensional surface, such as a sphere, to be embedded in a three-dimensional space in order to define its geometry. It's enough to consider measurements made entirely within that two-dimensional geometry, such as an intelligent ant, forever restricted to live on its surface, might make. The ant would know that the surface was curved by measuring that the sum of the internal angles of a large triangle differs from 180 degrees (as Gauss had done during his geodetic survey) or by measuring that the ratio between a large circumference and its radius differs from 2π. As a result of his study of surfaces, Gauss gave a precise mathematical meaning to the idea of curvature and a way of evaluating it. So-called Gaussian curvature is positive on the surface of a sphere, negative at every point on a saddle-shaped surface such as a hyperboloid, and zero for a plane. It thus determines whether a surface has elliptic (Riemannian) or hyperbolic geometry.

But Gauss didn't confine his thinking to a curved, two-dimensional surface floating in a flat, three-dimensional universe. In a letter to Ferdinand Schweikart in 1824, he dared to conceive that space itself is curved: "Indeed I have therefore from time to time in jest expressed the desire that Euclidean geometry would not be correct." This brilliant inspiration was to take root in the mind of Gauss's most talented apprentice.

Riemann extended Gauss's work to spaces of any number

of dimensions and put on a firm footing the type of non-Euclidean geometry that Gauss had hinted at, the kind known as elliptic geometry, in which there are no parallel lines and in which the angles of a triangle always add up to more than 180 degrees. He also generalized the notion of the shortest distance between two points. In Euclidean geometry this is simply a straight line. But step out of Euclid's domain and the quickest way to get from A to B involves a change of tack. The easiest way to grasp this idea is to think about traveling on Earth's surface, which isn't flat but (roughly) spherical–a special case of Riemann's elliptic geometry. To take a ship on the shortest route between two ports you sail, wherever possible, along an arc of a great circle–the circle that goes all the way around the Earth and on which both ports lie. Any such minimum-length path on a surface, the special case of which on a plane is a straight line, is called a geodesic, meaning "Earth divider."

In Euclidean geometry, the shortest distance between two points can be found using Pythagoras's theorem. What Riemann discovered was a more powerful, general form of Pythagoras's theorem that works on curved surfaces, even when the curvature is in more than two dimensions and varies from one place to another. In this looking-glass world of curved space, the familiar idea of distance is replaced by the broader concept of something called a metric, from the Greek for "measure," while curvature is similarly described by a more elaborate mathematical object. Gauss had found that the curvature in the neighborhood of a point of a specified two-dimensional geometry is given by a single number: the Gaussian curvature. Riemann showed that six numbers are needed to describe the curvature of a three-dimensional space at a given point and that twenty numbers at each point are required for a four-dimensional geometry: the twenty independent components of the so-called Riemann curvature tensor.

In his famous lecture of 1854, Riemann emphasized, as Gauss had done, that the truth about the space we live in can't be found by poring over 2,000-year-old books of Greek geometry. It has to come from physical experience. He pointed out that space could be highly irregular at very small distances and yet appear smooth on an everyday level. At very great distances, he also noted, a large-scale curvature of space might show up, perhaps even bending the universe into a closed system like a gigantic ball:

> "Space [in the large] if one ascribes to it a constant curvature, is necessarily finite, provided only that this curvature has a positive value, however small. . . . It is quite conceivable that the geometry of space in the very small does not satisfy the axioms of [Euclidean] geometry. . . . The properties which distinguish space from other conceivable triply extended magnitudes are only to be deduced from experience."

So far ahead of his time was Riemann that having arrived at his great mathematical description of space curvature, he began working on a unified theory of electromagnetism and gravitation in terms of it. Riemann grasped that forces might be nothing more nor less than a manifestation of the geometry of space. Flat beings on a wrinkly two-dimensional landscape, like that of a crumpled sheet of paper, would, when they tried to move around, experience what felt to them like gravitational effects. By analogy, he reasoned, forces in our world might best be explained in terms of warps in a higher dimension. And the effect would work both ways. If space told mass how to move, then space must itself–by the principle of action and reaction–be affected by mass.

The thirty-nine-year-old Riemann wrestled with these extraordinary possibilities in the summer of 1866, even as he lay dying of tuberculosis at Selasca on Lake Maggiore in Italy. He came close–astonishingly close–to a geometric theory of gravity half a century before Einstein, who later remarked of Riemann's contribution,

> Physicists were still far removed from such a way of thinking: space was still, for them, a rigid, homogeneous something, susceptible of no change or conditions. Only the genius of Riemann, solitary and uncomprehended, had already won its way by the middle of the last century to a new conception of space, in which space was deprived of its rigidity, and in which its power to take part in physical events was recognized as possible.

One major obstacle had blocked Riemann's further progress. He thought only of *space* and its topography. Einstein's great epiphany was that, in building a new theory of gravity, he also had to deal with *time*–with space-time and space-time curvature. But, to begin with, he didn't have the mathematical tools to do this. They existed; Einstein simply didn't know about them.

The Warp Factor

How do you stitch together countless tiny inertial patches to make a large, smoothly undulating quilt of curved space-time? As Einstein began thinking about this, he remembered that he'd studied Gauss's theory of surfaces in college. Suddenly he realized that the foundations of geometry had physical significance. To pursue the problem further, he contacted his old friend and talented mathematician Marcel Grossman. Einstein and Grossman had been students together at the ETH in Zurich; when Einstein skipped classes, he would often borrow Grossman's lecture notes. Einstein's overall mark at graduation

was a marginal 4.91 out of 6, which left him the only member of his class not to be offered a place in the ETH's physics department. He'd been written off, he said later, as "a pariah, discounted and little loved," virtually unemployable. Toward the end of 1901, still having found no permanent position, he wrote to Grossman explaining his plight. Fortunately, Grossman's father happened to be a friend of Friedrich Haller, the chief of the Swiss Patent Office, and so it was that Einstein got a desk job there despite Haller's opinion that he was "lacking in technical training."

Now, with gravity and curved space-time on his mind, Einstein once again turned to his trusted ally for help. Grossman had been appointed professor of descriptive geometry at the ETH in 1907 and had gone on to build a reputation as an outstanding teacher. He told Einstein of Riemann's work and of a subject called tensor calculus, especially the contributions made in the 1860s by Elwin Christoffel and more recently by Gregorio Ricci-Curbastro and Tullio Levi-Civita at the University of Padova in Italy. Abruptly thrown into a new and difficult field of math of which he'd previously been unaware, Einstein wrote, "[I]n all my life I have not labored nearly so hard, and I have become imbued with great respect for mathematics, the subtler part of which I had in my simple-mindedness regarded as pure luxury until now."

He had to learn about tensors–mathematical objects that behave in certain well-defined ways when you switch coordinate systems. (Vectors, for example, are a simple type of tensor.) Soon, it became clear to him that tensor calculus gave the perfect language for describing four-dimensional space-time. In 1913, Einstein and Grossman jointly published a paper in which they used the tensor calculus of Ricci-Curbastro and Levi-Civita to portray gravity in terms of a metric tensor (a tensor that gives a generalized way of measuring distance).[12]

But their theory was still far from complete. When Max Planck, the father of quantum mechanics, visited Einstein in 1913, and Einstein told him how things stood with his new scheme of gravity, Planck said, "As an older friend, I must advise you against it for in the first place you will not succeed, and even if you succeed no one will believe you."

For a while, it looked as if Planck might be proved right; not many scientists at the time thought Einstein was on the right track. Then, in October 1914, Einstein wrote a paper, nearly half of which was a treatise on tensors and differential geometry (the mathematics of surfaces). It proved to be a turning point because it led to a correspondence between Einstein and Levi-Civita in which the Italian pointed out technical errors in Einstein's analysis of tensors. Einstein was delighted by the exchange. Yet he continued to struggle with the equations that linked gravity with the geometry of space-time.

At the end of June 1915 Einstein spent a week at the University of Göttingen, where he lectured for six two-hour sessions on his (still incorrect) October 1914 version of what would become general relativity. Two of those present were colossi in the world of mathematics, David Hilbert and Felix Klein. "To my great joy," Einstein later recalled, "I succeeded in convincing Hilbert and Klein completely." Shortly after, Einstein and Hilbert began an intense exchange of letters on the outstanding problems in Einstein's theory. And now matters quickly came to a head. After chopping and changing the equations in his theory several times in the autumn of 1915–totally confusing his scientific colleagues in the process–Einstein made a monumental breakthrough. On November 18, 1915, he applied his new theory of gravitation to the old problem of Mercury's orbit and, lo and behold, found that it predicted, for the extra advance of the perihelion, exactly the 43 arc-seconds per century that astronomers had measured and that had foiled every other attempt at explanation. "For a few days," he remembered, "I was beside myself with joyous excitement."

To Hilbert, he wrote, "Today I am presenting to the [Prussian] Academy a paper in which I derive quantitatively out of general relativity, without any guiding hypothesis, the perihelion motion of Mercury discovered by Leverrier. No gravitation theory had achieved this until now." The Mercury figure was correct, but it was not yet the precise formulation. On November 25 Einstein submitted yet another paper, "The Field Equations of Gravitation," that at last contained the correct mathematical scaffolding of general relativity.[10]

There is a postscript: five days earlier, Hilbert had submitted a paper to a journal in Göttingen containing exactly the same field equations.[11] It has been suggested that Hilbert plagiarized Einstein, or perhaps vice versa; certainly, over those final frantic weeks before publication, each man came to know the other's thoughts well. But if the relationship between Einstein and Hilbert was strained for a while over the question of priority, it ended amicably enough, and Hilbert was able to write, "Every boy in the streets of Göttingen understands more about four-dimensional geometry than Einstein. Yet, in spite of that, Einstein did the work and not the mathematician."

Newton Eclipsed

Aristotle saw gravity as a property of matter; Newton considered it a somewhat mysterious force. But in general relativity, it's neither of these things. Gravity, in the brave new world of Einstein, is a manifestation of curvature in the geometry of space-time. As John Wheeler put it, "Matter tells space how to curve. Space tells matter how to move." (Here Wheeler is using "space" as shorthand for "space-time.") The Newtonian equivalent of this neat aphorism would be, "Matter tells matter how to move."

In many ways, general relativity turns our everyday notion of gravity on its head. Throw a ball straight up in the air, and a graph of its height versus time, seen through Newton's eyes, traces out a parabola. Einstein, however, recognized that a massive body–in this case, Earth–curves the coordinate system itself. Rather than following a curved path in a flat (Cartesian) coordinate system, the ball actually follows a minimum-distance path, or geodesic, in a curved coordinate system, returning to the thrower's hand at a later time because the geodesic leads it there.

This remarkable new view of things immediately removes two of the unanswered questions in Newtonian theory: How does gravity work? And why is the inertial mass of an object exactly equal to its gravitational mass? Einstein dismissed the first of these by showing that gravity isn't a force but simply a consequence of geometry. The second mystery also evaporates because, in general relativity, gravitational motion is seen as being nothing other than inertial motion in curved space-time. In other words, the equivalence of inertial and gravitational mass, which, under Newton, appears to be a curious and accidental fact, is seen in general relativity to be a necessary and unavoidable feature of the theory. In Einstein's scheme, inertial mass and gravitational mass aren't just accidentally numerically equal, they're ontologically identical.

Though seemingly counterintuitive when first encountered, general relativity is a beautiful piece of work–mathematically and conceptually. But beauty alone isn't enough to ensure survival. The acid test of any good scientific theory is whether the predictions it makes are borne out by experiment and observation. Chalk one up for general relativity for getting right the advance of the perihelion of Mercury. Then add credits for two other classic I-told-you-so's: one concerning the deflection of light rays from faraway stars that graze the sun, the other the phenomenon of gravitational redshift.

We saw earlier that Einstein was lucky to escape having his (erroneous) 1911 divination of how much light is bent by the sun's gravity put to the test. The new value that followed from the field equations of general relativity in 1915 was a factor of two larger at 1.74 arc-seconds. In 1919, the two British expeditions, one led by Eddington,[7] triumphantly confirmed this value to within the limits of experimental error, recording 1.98" ± 0.30" and 1.61" ± 0.30". As for a check on Einstein's prediction of a gravitational redshift, this had to wait much longer, until 1960, after his death. Very accurate (atomic) clocks were needed to test the premise that time really does slow down to the extent he foretold. When these clocks became available, general relativity was again completely vindicated.

Einstein's new vision of gravity superseded that of Newton. It explained what the older theory could not, in the most elegant way imaginable. It survived the classic tests of its accuracy. Next, scientists wanted to know what it could tell us about the universe that we hadn't previously been aware of most obviously, what it could tell us of the nature and origin of the universe itself.

8

Alpha and Omega

[W]hen you have eliminated the impossible, whatever remains, however improbable, must be the truth.

—SHERLOCK HOLMES

How did the universe begin? And how will it end? People have grappled with these great and terrifying questions since the birth of civilization.

According to the Cherokee, a mighty island once floated in a giant ocean. This island hung by four great ropes from the sky, which was solid rock. Because everything started out dark, the animals took the sun and put it in a path that bore it across the island from east to west each day. The animals and plants were told by the Great Spirit to stay awake for seven days and seven nights, but most could not. Those plants that succeeded, such as the pine and cedar, were rewarded by being allowed to remain green all year. The animals that stayed awake, such as the owl and the mountain lion, were given the ability to go about in the dark. Then people appeared.

In Norse mythology, the end of the world will be heralded by three winters without summers and a descent of human affairs into chaos. The wolves Skoll and Hati will swallow the sun and the moon, and the great wolf Fenris will run loose and kill Odin, who steps forth to fight him. Odin's son Vidar will avenge Odin by tearing Fenris apart. Many gods as well as all men and women, except two–Lif and Lifthrasir–who seek shelter under the branches of the great Yggdrasil, will die. The sky will fall into a pit of flames, and the earth will sink into the sea.

Our modern myths of beginnings and endings yield nothing in the drama department to these older tales. But in place of a pantheon of gods and monsters, science posits a single force that oversees the birth and death of all things. Gravity is the key to the origin and fate of stars, galaxies, and the cosmos as a whole–a fact that began to dawn in the wake of general relativity.

Constant Problems

The year 1917 was a bad one for Albert Einstein. Food was in short supply in Berlin because of World War I, he'd fallen ill with stomach and liver complaints (which would continue to dog him for the next four years), and his marriage to his first wife, Mileva, was on the rocks. To cap it all, he made what he later called "the biggest blunder" of his life: he missed predicting the expansion of the entire universe.

A year after announcing his new theory of gravity, Einstein published a paper called "Cosmological Considerations in the General Theory of Relativity."[11] It was the first-ever attempt to paint a mathematical picture of the universe as a whole, and, clearly, Einstein was pleased by his audacity. To his friend Paul

Ehrenfest, he wrote, "I have . . . again perpetrated something about gravitation theory which somewhat exposes me to the danger of being confined in a madhouse."

Einstein had dared to set his theory loose on all of space and time. Yet, uncharacteristically, he couldn't bring himself to accept what his equations told him. In their pure, unadulterated state, the field equations of general relativity said, in effect, that the universe couldn't be static. Either space-time, on a cosmic scale, is shrinking, or it is growing; it can't just stay the same. This certainly ran contrary to prevailing astronomical wisdom, which still strongly favored a universe that was stationary.

In fact, in 1917, it wasn't yet clear whether there was anything beyond our own galaxy. Einstein had never felt compelled to swim with the tide of orthodoxy in the past. So, why now? Perhaps his loss of nerve had something to do with his poor physical and mental state at the time. In any event, he made up his mind that the universe *was* stationary, and since his equations predicted otherwise, he felt it necessary to tamper with them–effectively, to slip in a fudge factor. He called this artificial correction the cosmological constant, denoted by the Greek capital letter Λ (lambda).

In some ways, the subsequent story of cosmology has been a saga of Lambda's ups and downs. From the outset, Einstein didn't like his added constant because it spoiled the elegant simplicity of the field equations he had struggled so hard to construct. His 1917 paper ends with the following: "It is to be emphasized, however, that a positive curvature of space is given by our results, even if the supplementary term [the cosmological constant] is not introduced. That term is necessary only for the purpose of making possible a quasi-static distribution of matter, as required by the fact of the small velocities of the stars."

Without Lambda, Einstein's equations insisted, a universe that started out stationary would immediately begin to shrink, slowly at first, but then with greater and greater urgency, drawn in upon itself by the mutual attraction of its gravity. Interpreted physically–not merely as a mathematical artifact–Lambda amounted to a *negative* pressure, a repulsion associated with the vacuum of space-time itself. Lambda was exactly the same everywhere in the universe, and its value was just such as to offset the tendency toward gravitational collapse at every point in space.

As far as Einstein was concerned, Lambda served its purpose. But its purpose, as originally conceived, was misguided. By the late 1920s Edwin Hubble and his assistant, Milton Humason, had accrued evidence, using the 100-inch telescope at Mount Wilson Observatory, that all the galaxies in space, apart from a handful that are very close to us, are flying apart. The universe is expanding–changing in size, just as Einstein's formulae had tried to tell him. Yet even before this discovery, it had emerged that Einstein had made a big mistake in forcing his model universe to be static.

In 1922 the Russian mathematician and astronomer Alexander Friedmann, director of the St. Petersburg Observatory, took up the problem of trying to describe the universe using general relativity. He realized that although a static universe was possible in theory–by precisely balancing gravity with a cosmological constant–such a universe would in practice be hopelessly unstable, like a pencil stood on end or a bowling ball perched at the top of a V-shaped mountain. The slightest irregularity in the distribution of matter would trigger a catastrophic expansion or contraction. It simply wouldn't do; you couldn't realistically insist that the cosmos was a special case that happened to be precariously balanced for all eternity. Friedmann allowed his mathematical universes the luxury of evolving, starting only from the assumptions that they contained matter throughout of constant density and were defined as a space-time of constant curvature. Three different cosmological models ensued, depending on whether the curvature was taken to be positive, negative, or zero.

A Friedmann universe with zero curvature, said to be flat, is just at the tipping point between collapse and expansion. The difference between this and Einstein's special stationary scenario, however, is that Friedmann didn't call upon a nonzero cosmological constant to hold gravity in check. A positively curved, or "closed" Friedmann universe is shaped analogously to a sphere and must ultimately collapse, even if, at some stage, it goes through a growth phase. By contrast, a negatively curved Friedmann universe, like the surface of a saddle, is permanently "open," its contents destined to move farther and farther apart for all time.

The factor that decides the overall cosmic shape, known as Omega, is the ratio of the actual density of mass and energy in the universe to the critical density at which the universe would be exactly flat. This means that in a closed universe, Omega is more than one, while in an open universe, it's less than one. A flat universe sees Omega equal to one exactly. Since Friedmann's day, a lot of detail has been added to our knowledge of the cosmos. We've learned about the Big Bang and the probable sequence of events that took place during it. We've probed the large-scale structure of the cosmos, including the clustering of galaxies. Yet the overarching question remains: Is the universe open (forever expanding), closed (doomed eventually to collapse), or flat (exactly in between)?

Einstein's initial reaction to Friedmann's dynamical models wasn't overly enthusiastic: "The results obtained in the work cited regarding a non-stationary universe seem suspicious to me."

Three years later, Friedmann died, at the age of thirty-seven, before the significance of his work was recognized. The attitude of scientists at the time to suggestions of anything other than the unchanging cosmos of the status quo is summed up by the English physicist J. J. Thomson (discoverer of the electron): "We have Einstein's space . . . expanding universes, contracting universes, vibrating universes, mysterious universes. In fact the pure mathematician may create universes just by writing down an equation, and indeed if he is an individualist he can have a universe of his own."

Such criticisms would fade with Hubble's revelation of the fleeing galaxies. In January 1931, Einstein and his second wife, Elsa, were guests of Hubble at Mount Wilson Observatory. The fifty-one-year-old Einstein scrambled excitedly over the framework of the great telescope like a kid at a playground while the tour guide explained that one of the tasks of the instrument was to determine the shape and extent of the universe. "Oh," replied Elsa, "my husband does that on the back of an old envelope." But seeing is believing, and doubtless Einstein was also impressed by what he saw through the eyepiece that evening. During the following year, he finally abandoned his cosmological constant. It was to the Russian physicist George Gamow, one-time student of Friedmann, that Einstein later admitted that introducing Lambda had been a blunder.

Yet that confession, appropriate though it may have seemed at the time, has turned out to be premature. As we'll see in chapter 11, the cosmological constant has refused to go quietly into retirement. It resurfaced in the 1970s in a remarkable new guise. And, still more recently, it has claimed center stage again–this time in a form that Einstein would have found strangely familiar.

Gravity, and its dark nemesis, hold the key to the future of everything. In the same way, on a smaller scale, gravity and the forces that oppose it determine the destiny of stars and the great cities into which they congregate.

A Fatal Attraction

Like people and universes, stars are born, mature, grow old, and die. The manner of their demise, however, varies enormously. Some–the stellar lightweights (red dwarfs and their kin)–end with a long, slow fade into obscurity. Others inflate extravagantly to become bloated red giants before more or less gently casting off their outer parts, at the same time as the old stellar core shrinks to become a planet-sized ball of hot, inert matter. Such is to be the sun's destiny several billion years from now. For stars of greater mass, however, a far more spectacular fate lies in store. A heavyweight star dies in a blaze of glory, an explosion called a supernova, that can briefly outshine an entire galaxy of 100 billion suns. All that's left behind is an incredibly compressed object resembling an atomic nucleus as wide as Chicago. Alternatively, the residue might fold in upon itself still further, to something even more bizarre. It might disappear down a bottomless pit in space-time–a black hole–out of which not even light can escape.

The term *black hole* was coined as recently as 1967 by John Wheeler. But the concept goes back much further, to 1783, when it first occurred to John Michell, a small, chubby English vicar, whom we met in chapter 6 as the inventor of the torsion balance used by Henry Cavendish to "weigh the world." Born in Nottinghamshire in 1724, three years before Isaac Newton died, Michell was one of the most remarkable unsung polymaths of his age–a fantastically creative Renaissance man. He went to Queen's College, Cambridge, where, in due course, he taught arithmetic, geometry, theology, Greek, Hebrew, and philosophy, all while carrying the official title of professor of geology. Wearing his geologist's hat, he was the first person to suggest that earthquakes travel in waves, and he thus founded the field of seismology. He also proved the inverse-square law of magnetism (using the same balance that he later adapted to measure gravitational forces) and tried to figure out how much pressure is exerted by sunlight.

In 1767, at the age of forty-three, Michell left Cambridge for the quieter, but surprisingly well-paid, life as rector of the church of St. Michael and All Angels at Thornhill, near Dewsbury, Yorkshire–hardly the hub of the intellectual universe. Yet here in this academic backwater, the portly pastor continued his scientific investigations and drew in a wide circle of influential friends. The rectory at Thornhill was a fine old house with a beautiful lawn and ancient trees, and it needs no great leap of the imagination to picture Michell and his colleagues strolling together through the grounds or sipping postprandial port in the dining room and discussing the latest philosophical ideas–a northern version of the celebrated Lunar Society of Birmingham. Joseph Priestley, the prominent chemist (discoverer of oxygen) and fellow clergyman, used to ride his horse over from the nearby city of Leeds, as probably, too, did the civil engineer John Smeaton. Cavendish, who eventually inherited Michell's balance, was another frequent visitor. And Michell liked to play the violin with William Herschel, the Hanoverian musician who came to England and, perhaps partly inspired by Michell, went on to become a world-renowned astronomer and discoverer of Uranus.

Of all his varied interests, none inspired Michell more than astronomy. In his first year at Thornhill, he asked a question about the famous Pleiades, or Seven Sisters–the well-known fairy-dust of stars in Taurus, so prominent in northern winter skies. If stars were scattered randomly across the heavens, wondered Michell, what were the odds of a grouping as close as the Pleiades appearing by chance? His answer: one part in half a million–so improbable, he concluded, that the Pleiades must be a stellar system in its own right, a genuine physical clustering of stars. It was to be the first known application of statistics to astronomy.

Later, Michell wrote a letter to Cavendish that was eventually published in *Transactions of the Royal Society* in 1784, with the bewildering title (not untypical for its time) "On the Means of Discovering the Distance, Magnitude, etc. of the Fixed Stars, in Consequence of the Diminution of the Velocity of Their Light, in Case Such a Diminution Should Be Found to Take Place in Any of Them, and Such Other Data Should Be Procured from Observations, As Would Be Farther Necessary for That Purpose."[20] His idea was to figure out stellar distances by measuring the speed of light coming from stars–the farther away the star, he thought, the slower its light would be moving. A better quality of prism would be needed, he realized, for his measurements to be carried out, but such a prism was certainly feasible.

Somewhere in the middle of the paper, he noted that if a star had the same density as the sun but a radius 500 times bigger, then a body falling toward it from infinitely far away would move faster than light, "and consequently supposing light to be attracted by the same force in proportion to its *vis inertiae*, all light emitted from such a body would be made to return towards it, by its own proper gravity." Michell called his hypothetical lightless creature a "dark star" and surmised, with extraordinary foresight, that its detection might be possible if it lay within a binary system–a pair of stars revolving around their common center of gravity–by observing the motion of the companion. The concept of an object with a gravity pull compelling enough to prevent light from escaping was born.

Michell's ideas about dark stars were echoed the following decade in France by Pierre Laplace. In the first two editions of his popular astronomy guide *Exposition du Système du Monde* (Exposition of the System of the World), published in 1796 and 1799, Laplace talks about a super-sun-sized dark star similar to that of Michell's, commenting, "[I]t is therefore possible that the greatest luminous bodies in the universe are on this account invisible." But by the time of the third edition in 1808, he'd dropped the subject.

Michell had grasped the basic physics of what we now call black holes. But, not surprisingly, considering how early he was on the scene, he got some of the details wrong. With hindsight it's clear that the prism experiment he proposed, and which was actually carried out by the astronomer Neville Maskelyne, was doomed to fail. Both of Einstein's relativity theories, special and general, rest on the assumption that light from all sources has the same velocity when it reaches us. The reason it can be "slowed down" is the distortion that gravity produces in space and time.

We also know that light has many wavelike properties, which aren't taken into account in the Newtonian particle, or "corpuscular," theory of light that Michell used. A major reason that Michell's speculations were sidelined in the nineteenth century was the overthrow of Newton's particle theory of light by the wave theory devised by Thomas Young in 1801. It wasn't until the rise of quantum theory that the concept of light particles, or photons, became scientifically respectable again. Since light waves were thought to be unaffected by gravitation, interest in hypothetical "dark stars" went away–at least for a while.

Event Horizon

Within weeks of the publication of his general theory of relativity, Einstein received a paper from Karl Schwarzschild, a German astrophysicist and director of the Potsdam Observatory.[34]

Away from his academic post because of World War I, Schwarzschild was serving with the German army on the Russian front but had been laid low by a rare metabolic illness that would soon prove fatal. Hospitalized and bedridden, Schwarzschild turned his thoughts to the astronomical implications of Einstein's new theory. He wondered what the relativistic field equations would have to say about the gravitational field around a point mass–a mass concentrated at a single, infinitely small point in space. Although no such object really exists, it isn't a bad approximation to a star, providing that the star is on its own, spherically symmetric, and nonrotating. So, Schwarzschild's solution to Einstein's equations, known as Schwarzschild geometry, gives a fair picture of how space-time is curved in the vicinity of a star.

Included in this mathematical description is a special distance that has come to be called the Schwarzschild radius, or event horizon–the distance at which the escape velocity would equal the velocity of light. Nothing that's inside the event horizon can ever get out, because to do so, it would have to break the light barrier. From this it follows that if the entire mass of an object lies inside its event horizon, then it would be cut off from the outside world; it would be a dark star, or black hole. Any mass becomes a black hole if it's shrunk small enough. To make a black hole from the sun, for example, you would have to squash all of its matter into a ball about a mile and a half across. The Earth would become a black hole if it were squeezed to roughly the size of a marble. A black hole's size–its Schwarzschild radius–is simply proportional to its mass. So, if you double the mass, you also double the size.

Despite Schwarzschild's publication of his solution, no one took black holes seriously at the time. There didn't seem any way they could possibly form in the real universe, and so any theoretical talk of them appeared to have no application. That was about to change, however. In the 1920s, just as the success of the theory of relativity led to an explosion of interest in it, scientists started to figure out how stars worked. Arthur Eddington and others realized that stars, such as the sun, produce light and heat by nuclear fusion in their cores. They also reasoned that the pressure of radiation produced in the stellar interior is what prevents a star from collapsing under its own gravity. Once the central nuclear fuel is used up, however, it is clear that something has to give. Researchers then began to ask questions about the later stages of stellar evolution, when the outward pressure of light and heat fails and the means of resisting gravity disappears.

In 1931, the Indian-born astrophysicist Subrahmanyan Chandrasekhar, who spent much of his career at the University of Chicago, became the first person to correctly apply the equations of relativity to the interior of a so-called white dwarf.[5] A white dwarf is the kind of object that a star like the sun, or one that is somewhat more massive, will turn into when all of its useful nuclear fuel is exhausted. The old stellar core collapses to a ball approximately the size of Earth, so dense that a piece of it no bigger than a sugar cube would outweigh a hippopotamus. It is then prevented from shrinking still smaller by the refusal of the electrons it contains to become any more crowded. This so-called electron degeneracy pressure, a quantum effect, is what supports the star in its wizened state from being compressed by gravity any further. What Chandrasekhar showed, however, is that stabilization at the white dwarf stage can happen only if the star has clung on to less than about 1.4 solar masses of material. Any more than this, and gravity will overwhelm the electrons' dislike of living in one another's pockets, or, strictly speaking, one another's energy states, forcing the star to continue its inward gravitational plunge.

Beyond the white dwarf stage lies that of the neutron star, a type of object first described in 1934 by the astronomers Walter Baade and Fritz Zwicky. In this state, most of the particles that made up the old star are jammed together so hard that they combine to form a sphere of almost pure neutrons. The result is a ball, 10 to 15 miles in diameter, with an almost inconceivable density of about a billion tons per cubic inch. Supporting the crushed star now is another quantum effect known as neutron degeneracy pressure. In time, astronomers came to realize that fast-spinning, highly magnetized neutron stars can be powerful emitters of radiation, mostly in the form of radio waves. These waves are sent out in the manner of a lighthouse beam, so that if we happen to lie along the line of sight of the beam we're treated to a fast, regular succession of radio (and, in some rare cases, also light) pulses. The first source of this kind was discovered in 1967 and quickly became known as a pulsar.

Five years after Baade and Zwicky laid down the basic science of neutron stars, J. Robert Oppenheimer, best known as the leader of the Manhattan Project, and therefore as the father of the atom bomb, and his colleague George Volkoff found that these objects, too, have an upper mass limit. If a dead star has more than about three times the mass of the sun, they demonstrated, there is nothing in the universe that can prevent gravity from crushing the star to the ultimate condensed state–a black hole.[26]

Even with this knowledge, scientists didn't give the idea of black holes much attention. World War II loomed large in everyone's mind, and, besides, even if black holes existed, there seemed to be no way we could ever observe them. How can you study something that emits nothing?

On the theoretical front another breakthrough came in 1963, thanks to the New Zealand physicist Roy Kerr. He found out how to apply Einstein's field equations to the space-time around *spinning* stars–a much more realistic scenario than the static case analyzed by Schwarzschild. Kerr's initial announcement came in the form of a paper just a-page-and-a-half long.[17] He followed this up, however, a couple of years later with a much more detailed description of his spinning solution that contained a very interesting feature–especially if you're thinking of venturing inside a black hole sometime.

Truthfully, black hole travel isn't to be recommended. Get too close to one of these things, and (with certain intriguing exceptions, as we'll see in chapter 13) there's every chance you'll be torn limb from limb, and atom from atom, by tidal forces. If you somehow manage to survive the passage through the event horizon, then you're still in an awful predicament because there's no way back out into the universe at large. You're stuck inside the black hole. And matters can only go downhill from there.

Within the black hole, space-time is extraordinarily distorted, so much so that the coordinates describing radial distance and time switch roles: space assumes the part of time, and vice versa. One consequence of this is that you can no more stop yourself from getting closer and closer to the center of a black hole than, under normal circumstances, you can avoid moving toward the future. Eventually, you're bound to hit the middle. Fire your retrorockets as hard you like in an effort to prevent yourself going deeper into the black hole, but it's a waste of time (or maybe space). Trying to avoid the center of a black hole once you've crossed the event horizon is like trying to avoid next Friday; it can't be done. You're doomed in fairly short order to have an unpleasant encounter with the phenomenon that lurks in the middle of this space-time prison. Called a singularity, it is a place where all the readings of density, pressure, and space-time curvature go off the scale. Upon meeting it, whatever of you is left will be squeezed out of existence.

At least, that's the situation inside a Schwarzschild black hole–one that's not spinning. Kerr found that if rotation is added to the mix, the situation changes: the black hole doesn't end up as a dot of zero size. Instead, it takes the form of a spinning ring, which is kept from collapsing by centrifugal forces. Although black hole excursions would probably still count as the most extreme of adventure vacations–no return ticket necessary–Kerr's solution does offer a way, at least in principle, of avoiding the central singularity and even of navigating through the black hole to some other place in space and time, possibly even to another universe.

Acceptance

While theorists toyed with such exotica and gave science-fiction writers a field day, observational astronomers finally began to catch black-hole fever. The 1960s saw the discovery of quasars and other vastly energetic events taking place in the cores of faraway galaxies that would eventually be put down to the presence of supermassive black holes. These dark galactic hearts–even our own modest Milky Way is strongly suspected of harboring one–can out-mass 100 million suns and feast on any debris, including whole stars, that venture too near to their gravitational domain. Captured material enters a swirling vortex around the behemoth, spinning faster and faster, like water round a plughole, and heating up, because of friction, as it slips and slides against other matter, to millions of degrees. At these temperatures, the tortured gas begins to emit electromagnetic radiation of all wavelengths up to and including those of X-rays. So, by searching for telltale flickering of a source at X-ray wavelengths, astronomers have a way of looking for black holes and the fiery whirlpools, known as accretion disks, that form around them if a supply of matter is available.

The launch of X-ray detectors and telescopes into orbit provided our earliest views of the universe at these short wavelengths, which, fortunately for our health, are absorbed by Earth's atmosphere. One of the first great discoveries, in 1972, was of X-rays coming from a source known as Cygnus X-1. Observations in ordinary light of this object, in tandem with X-ray measurements, revealed it to be a binary system in which a giant blue star and an unseen companion circle around each other every five-and-a-half days. The visible star, catalogued as HDE 226868, is a so-called O-type star–one of the biggest and brightest members of the stellar fraternity. Its mass, unless astronomers are very much mistaken, could hardly be much less than about thirty times that of the sun. Orbital statistics would imply that its partner weighs in at roughly ten solar masses.

Oppenheimer had shown before World War II that there could be only one candidate for a dark star as heavy as this. Cygnus X-1 almost certainly contains a black hole. We say, "almost certainly" because we have to admit that there's still a bit of room for doubt. Evidence for black holes, due to the nature of these beasts, is circumstantial. Maybe we're wrong about the mass of HDE 226868, and, despite all our understanding of stars and their evolution, this stellar luminary is much lighter than we suspect. If so, it would mean that its companion is lighter too–perhaps light enough to rank as "only" a neutron star. But there are very few professional astronomers today who believe this. Fly in a spacecraft close to Cygnus X-1, and you would almost certainly see a stream of bright gas heading from the blue giant star across space to a nearby vortex centered on a black hole.

All the data we have suggest, too, that supermassive black holes at the center of large galaxies are ubiquitous. If a supply of stellar food is available for these gravitational monsters, they–or, rather, their accretion disks–flare up and glow brilliantly across the spectrum as quasars or their "active galaxy" cousins, the Seyferts and radio galaxies. If, as a galaxy matures, it runs low on tidbits for the monster that inhabits its core, then the quasar light–the extra luminosity across the spectrum, over and above that supplied by the stellar population–dims. At any time, however, a galaxy such as our own has the potential to flare up to unusual brilliance if a fresh supply of material, say a wayward star cluster, becomes available to the central black hole.

In the 1970s, the British astrophysicist Stephen Hawking suggested a possible third species of black hole in the universe–miniature black holes, large numbers of which might have been formed at the time of the Big Bang. Further, said Hawking, we might witness these Lilliputian pits in space-time exploding. This seems to contradict what we said earlier about nothing being able to escape from a black hole once it's crossed the threshold of the event horizon. But Hawking was able to show that there's a subtle quantum effect by which black holes can indeed leak stuff back into the space around them. This seepage, known as Hawking radiation, is expected to occur from any black hole, large or small, but it gathers pace as the hole evaporates. Finally, as its mass drops to next to nothing, the black hole erupts in a final flourish of gamma rays. Although no such events have yet been recorded, scientists have no reason to doubt that Hawking radiation is real and that, as a result, any black hole must eventually relinquish its contents, even if it takes a trillion, trillion, trillion, trillion years or more.

One final thought: if the universe is closed–in other words, if the whole of space-time curves back on itself like the surface of a ball–then we're driven to an unsettling conclusion, namely, that we actually live within a black hole of heroic proportions. If such is the case, does this greatest of all black holes also emit Hawking radiation? And if it does, where in heaven's name does all this radiation go?

9

The Ripples of Space

Some things have to be believed to be seen.

—RALPH HODGSON

Some two thousand light-years away, in the constellation of Aquila the Eagle, lies one of the strangest objects in our galaxy. Two stars are silently whirling around each other in less time than a nine-to-five stint at the office, their point of closest approach little more than twice the distance from Earth to the moon. These partners, dancing so close and frenetically, are no ordinary stars, however. Each is so dense that a pinhead-sized speck of its matter would outweigh a battleship. They are neutron stars. And, crucially, one of them is also a pulsar–its bright, insistent lighthouse-beam of radio waves sweeping across our line of sight every 59 thousandths of a second. That precise beacon helped its discoverers win the 1993 Nobel Prize for Physics for it confirmed a remarkable prediction of Einstein's theory of gravity made a lifetime earlier.

Bangs and Whispers

In general relativity, massive objects affect the way that space-time curves. Picture space-time as the elastic skin of a trampoline. A person standing on the trampoline creates a hollow in the rubber sheet, just as a mass in space-time causes a dip to form in the space-time around it. If the person bounces up and down on the trampoline, she causes vibrations to travel out across the sheet. In the same way, says general relativity, a massive object that suddenly shifts will create ripples in the fabric of space-time–ripples known as gravitational waves.

In a sense, the idea of gravitational waves was implicit in Einstein's 1905 special theory of relativity, with its finite limiting speed for the transfer of mass, energy, or information. No effects, including those of gravity, he insisted, can get from one place to another faster than the speed of light. But the exact equations describing gravitational waves weren't revealed until general relativity came along. In 1916, and in more detail in a 1918 paper, Einstein showed that when a mass accelerates–in other words, changes its state of motion–it can't help but give rise to time-varying gravitational fields that travel away from the source at light-speed as undulations in the surface of space-time.

Anything with mass, or any collection of things with mass, spawns gravitational waves when it changes shape, spatial arrangement, or velocity. Think of these waves coming about in the same way that electromagnetic waves do. An electrically charged moving object gives off electromagnetic waves in proportion to its charge and acceleration. Likewise, a moving mass generates gravitational waves in proportion to its mass and acceleration. There's just one important difference. Newton's third law of motion says that the acceleration of a mass in one direction must be accompanied by an acceleration of another mass in another direction, with momentum (mass × velocity) in both directions being equal. This means that the other mass generates gravitational waves as well, so that the waves of the two masses tend to cancel out. Because the masses aren't in the same place, however, the cancellation is never perfect. The amount of gravitational radiation that gets away depends on the arrangement of the masses, measured by what's called the *quadrupole moment.* A totally symmetrical object, like a soccer ball, has zero quadrupole moment, whereas something very unsymmetrical, like a football, has a large quadrupole moment, at least for rotation around its short axis.

If a mass undergoes some sudden change, the resulting gravitational waves will take the form of a short pulse, much like the first big ripple produced after dropping a rock into a still pond. In the case of a periodic change, the wave will be sustained, like the carrier wave for a broadcast radio signal. In either case, the amplitude, or height, of the wave will steadily fall as the distance from the source increases.

Gravitational waves are *transverse,* as are light and water waves, which means that they vibrate at right angles to the direction in which they're traveling. To appreciate how this affects their interaction with matter, suppose four masses are arranged at the four points of a compass in the horizontal plane and a gravitational wave passes through the plane from above. At one point, the distance between the north-south masses is decreased and the distance between the east-west masses is increased. A half wavelength later, the reverse is true. If the gravitational wave passes through the masses *along* the direction of the plane, say in the east-west direction, it has no effect on the masses in the direction of its motion. The east-west masses remain fixed in position while the distance between the north-south masses increases and decreases as the wave passes through the plane.

If general relativity is to be believed, the universe should be awash with gravitational waves. Yet Einstein never took them seriously. He tended to think of them as mere mathematical artifacts–things that were possible in theory but that, in reality, would never be strong enough to have measurable effects. It was a view shared by most of his colleagues, among them Arthur Eddington, whose eclipse measurements gave general relativity such a powerful boost. Derisively, Eddington commented, "Gravitational waves propagate at the speed of thought."

But what if they existed *and* could be picked up and recorded? Then they might open up a whole new window on the cosmos. In the 1950s, the first giant radio telescopes began giving astronomers an astonishing view of the universe beyond that available in ordinary light. Around the same time, theoreticians managed to show that gravitational waves are, at least in principle, detectable.

In principle. It's important to get the scale of the observational problem in perspective. Imagine that several thousand light-years away some stellar cataclysm occurs that generates the gravitational equivalent of a tsunami. By the time that once-mighty wave has reached our stretch of the galactic shore, its influence on space-time, as it passes by, will be to cause a distance as large as that between Earth and the nearest star (after the sun)–about 40 trillion miles–to alter temporarily by about the width of a human hair. An instrument sensitive enough to detect that would be able to sense a change in the Earth-sun distance of about the width of an atom: roughly a millionth the diameter of a proton per meter. Any way you slice it, it's asking a lot of our engineering ingenuity.

The detection problem goes back to the feebleness of

gravity. Even when large, sudden movements of huge amounts of matter are involved, the waves produced are bound to be fantastically weak by the time they've made their way to us across many light-years of space. A gravitational wave arriving on Earth will alternately stretch and shrink distances, but on a mind-numbingly small scale.

Yet the promise of what gravitational wave astronomy might tell us about the universe is too great to allow technical difficulties to stand in the way if we can help it. The most powerful gravitational waves will be produced by the most violent events. Many of these are expected to involve matter in its most condensed state, in the form of black holes and neutron stars—bizarre denizens of the cosmos unknown in Einstein's day. Great surges of gravitational radiation are thought likely to come from the implosion of a stellar core at the heart of a supernova to form a black hole or a neutron star, by the swallowing of a neutron star by a black hole, or by the collision and merging of two black holes. Observing gravitational waves could give us an independent means of estimating cosmological distances and of peering back to the earliest moments when space and time came into being. They might also reveal phenomena of which we previously had no inkling.

Almost all our information about the cosmos at present comes in the guise of electromagnetic waves—visible, radio, infrared, ultraviolet, X-ray, and gamma-ray. A smattering arrive by way of cosmic rays and neutrinos, but that's about it. Gravitational waves offer an entirely new vista on what's out there, especially at the high-energy, fast-changing end of the phenomenological scale. And, although their detection poses enormous challenges, they do have at least one advantage over their electromagnetic counterparts: whereas light, radio waves, and so forth are affected by intervening matter (light is strongly absorbed by interstellar dust, for example), gravitational waves are expected to pass unchanged through any material in their path and so be able to carry signals with absolute clarity across the vast reaches of space.

Listening Posts

The late 1960s saw the first gravitational wave detector take shape at the University of Maryland in College Park. Its creator, Joseph Weber, then caused a minor sensation by claiming a couple of years later that his instrument had come up with the goods.

Weber's detector used two huge (1.5-ton) solid aluminum cylinders to which piezoelectric crystals were wired. These two bars were placed at separate locations in order to sift out random noise and other false readings, the idea being that only signals that affected both bars would be recorded. If a suitable gravitational wave came along, it would trigger the so-called fundamental longitudinal mode of the cylinder, whose natural frequency was around 1,600 hertz (cycles per second), causing the cylinder to deform like a rubber stopper that's alternately stretched and squeezed along its lengthwise axis. This frequency was chosen because it's in the range that astrophysicists had predicted for the short pulses of gravitational waves expected from supernovae. In Weber's setup, the piezoelectric crystals bonded to the surface of the cylinder were intended to respond to any deformations by producing electrical voltages that could be read out.

Not long after firing up his rig, Weber saw what he took to be strong evidence for gravitational waves; the number of periodic signals detected, he reported, far outweighed the amount of noise, and, when the detector was pointed for optimum reception of waves from the center of the galaxy, the recorded observations peaked.

Weber's claims, however, ran into a wall of skepticism. For one thing, it seemed that his equipment was nowhere near sensitive enough to be picking up the number of hits he was talking about. For example: The strongest gravitational waves coming from a binary system in which there are two collapsed objects, such as neutron stars or black holes, orbiting close to each other, would distort a gravitational wave detector by about one part in 10^{20} (100 million trillion). The supernova that occurred in the Large Magellanic Cloud (a satellite galaxy of our own) in 1987–the nearest, brightest such event in more than three centuries–might have managed a distortion of one part in 10^{19}. And an even closer supernova in the disk of the Milky Way would push that distortion up by another order of magnitude to about one part in 10^{18}. Yet one part in 10^{18} was the supposed limit of sensitivity of Weber's apparatus. His positive readings couldn't possibly have been due to nearby supernovae because none had been confirmed by other means; in fact, the last recorded supernova in the Milky Way was the one seen by Kepler in 1604. Either gravitational waves, in general, were at least a thousand times more powerful than anyone had predicted, enabling Weber to detect them coming from sources less violent than supernovae, or something was wrong with his experiment.

In the end, the latter turned out to be the case. Weber's machinery was riddled with so much noise that any genuine gravitational wave signal would have been hopelessly masked. The biggest source of this background chatter was the incessant thermal vibrations of the atoms in the metal cylinders. These alone were enough to cause a longitudinal oscillation a hundred times stronger than any that might stem from a gravitational wave.

More experiments followed around the world using the so-called resonant antenna method that Weber had pioneered. All these strived to improve sensitivity and cut noise in various ways, most importantly by cooling the test masses to temperatures close to absolute zero so that thermal motions were kept to a minimum. Another problem was the noise of the electrical transducers and amplifiers used to process the gravitational wave signal. Modern detectors have done away with piezo crystals in favor of cavity resonators or coils whose electric properties are changed by the deformation of the test mass. To boost the signal noiselessly, they use SQUIDs (superconducting quantum interference devices) in which quantum effects do the job of amplification. And, in the latest bar detectors, the whole test mass is suspended as a pendulum to isolate it as much as possible from seismic and other external disturbances. Aided by all these tricks, today's resonant detectors can sense deformations of 10^{-19} meters, comfortably giving them the sensitivity needed to detect the strongest of gravitational waves. So far, they've heard nothing. But if there's another supernova in our galactic neighborhood, like the one in 1987, or better still in the Milky Way itself, that might change. Efforts, too, are continuing to improve the sensitivity of resonance antennae.

At the same time, scientists have realized there's another way to bring gravitational waves into view. Known as laser interferometry, it exploits an arrangement similar to that of the famous Michelson-Morley experiment described in chapter 7. A laser interferometer gravitational wave detector (LGD) doesn't measure the distortion of masses as such. Instead, using a laser beam, it looks for a change in length of a pair of long pathways at right angles to each other.

The idea of applying laser interferometry to the problem of detecting gravitational waves was dreamed up in the late 1960s by Ranier Weiss of the Massachusetts Institute of Technology (MIT). A decade later, encouraged by the physicist Kip Thorne (who, as a student at Princeton, had been a fan of Weber's work), a well-funded research effort on LGDs sprang up at the California Institute of Technology (Caltech). Then, in 1983, the Caltech and MIT groups got together and began what evolved into the biggest effort to date to detect gravitational waves. Called LIGO (the Laser Interferometer Gravitational-Wave Observatory), it was built at a cost of $365 million and kicked off its scientific work in earnest in 2005.

LIGO consists of two identical facilities: one at Hanford, on the plains of south-central Washington State, and the other at Livingston, Louisiana, in a forest near Baton Rouge. At each location is a strange L-shaped building with two skinny arms two-and-a-half miles long. At the vertex of the L is the nerve center of the operation, where a flashlight-sized, ultrastable laser beam is split and sent down each arm inside a four-foot-diameter, stainless steel vacuum tube. At the far end of the arm, a delicately suspended mirror reflects back the laser light, and the two returning beams are recombined and fed to a photodetector. A passing gravitational wave will slightly compress one tube while simultaneously stretching the other, causing one mirror to move slightly forward and the other slightly back. The effect of that tiny movement is to cause a slight phase shift in the laser light traveling along the two arms; it is this phase shift that the detectors are designed to measure. Having two installations in different parts of the country helps scientists see past the extraneous noise in the system. Events detected by LIGO will only be considered potentially valid if they happen within 10 milliseconds of each other–the maximum time taken by a gravitational wave to travel the 1,890 miles between Hanford and Livingston.

In 2005 LIGO was still being put through its paces, and researchers weren't concerned that it had yet to record its first positive signal. Improvements to the system, begun in 2005, will result in the LIGO 2 configuration, which is widely expected to achieve the long-awaited detection of gravitational waves. Featuring more powerful lasers, greater immunity from extraneous vibrations, and more accurate mirrors made of single-crystal sapphire, LIGO 2 will be fifteen times more sensitive than its predecessor; meanwhile, researchers are already considering ideas, such as mirrors that are cryogenically cooled, for LIGO 3.

One of the big advantages of an interferometer over a bar-type detector is that it can search for gravitational waves with widely differing frequencies. A bar detector has to be specially constructed, much like a tuning fork, for each frequency. Ultimately, though, the goal is to link all the world's major gravitational wave observatories, of whatever type, into a global network, so that as much information as possible can be extracted from any signals received. Comparing the times when a wave passes different detectors, for example, will give information about the direction of the source in thc sky. Accurate timing, strength, and frequency data will also help shed light on the physical processes that created the waves.

Ground-based gravitational wave detectors are expected to make huge strides in the early twenty-first century. These facilities are inevitably subject to vibrations of Earth itself, however, rendering them useless for picking up low-frequency waves. This type of gravitational wave is thought to be created when a supermassive black hole at the heart of a galaxy swallows a star. So, if we hope to be able to "see" the space-time ripples from such an event, we need to put our gear where no planetary rumblings can affect it–in space.

Enter LISA (the Laser Interferometer Space Antenna). This project, being developed jointly by NASA and the European Space Agency, will involve placing three identical satellites in orbit in the configuration of a triangle with sides 3 million miles long. Each LISA satellite will look like a ring with a Y-shaped core, and each will contain a laser and telescope assembly to allow it to link up with its two partners. The test masses will be cubes one and a half inches on a side, floating freely inside the spacecraft. A prototype system is expected to be launched first, followed by the full LISA array in about 2011.

The Odd Couple

Gravitational waves have yet to be recorded directly. But scientists no longer seriously doubt they exist. Much of that optimism stems from a discovery made in 1974 by two radio astronomers at the University of Massachusetts at Amherst, Joseph Taylor and his graduate student Russell Hulse (both now at Princeton).[15] Taylor and Hulse had been using the giant Arecibo radio telescope in Puerto Rico to search for new cosmic radio sources when they picked up a highly regular pulsed signal coming from a point in the sky just a few degrees away from the bright star Altair. The little radio beacon pulsed seventeen times a second, as regular as clockwork. Taylor and Hulse knew right away that they'd found a new pulsar (a fast-spinning, highly magnetized neutron star)—an object that would be catalogued as PSR 1913+16.

Another pulsar added to the inventory isn't in itself such big news; over 700 have been identified since the first came to light in 1967. But as the two astronomers studied their new discovery more closely, they uncovered something very unusual indeed: there was a systematic variation in the arrival time of the pulses. Sometimes, the pulses came a little sooner than expected—up to three seconds sooner. Sometimes they came a little later, by up to the same amount. These variations happened in a smooth and repetitive manner, with a period of 7.75 hours. Taylor and Hulse now knew they had captured a much rarer animal. This was no pulsar wandering through the galaxy alone. It had a companion, a stellar partner, and the two stars were engaged a rapid pas de deux.

Seen in the context of a binary system, the changing arrival times of the pulses make sense. When the pulsar is on the side of its orbit closest to Earth, the pulses arrive earlier that they do when the pulsar is on the side farthest from us. The difference of three seconds corresponds to the time taken for radio waves to travel the one-and-a-half-million miles from one side of the orbit to the other. Further observations showed the orbits of the two stars around their common center of gravity to be quite eccentric, or stretched out. At minimum separation, known as periastron, the pulsar and its companion are only about one solar radius apart; when furthest apart, at apastron, slightly more than two suns could be fitted in the gap between them.

The beauty of a binary system is that the mutual motions of the two components reveal the secret of their masses. Very quickly, Taylor and Hulse were able to establish that the pulsar and its dance partner had an almost identical mass: about 1.4 times that of the sun. Since neither star showed up on optical pictures, the conclusion was inevitable–and extraordinary. The pulsar's companion was also a collapsed object–another neutron star. Taylor and Hulse had discovered the first binary neutron star system.

Astronomers had long speculated that such systems must exist. They also knew that if two dense, heavyweight objects were found, whipping around each other in a tight, fast orbit, they'd serve as an ideal natural laboratory for testing various predictions of general relativity. The existence of the pulsar was a godsend because the pulsing of its radio emission was like the ticks of an accurate clock. The discoverers of PSR 1913+16 had uncovered a gravitational goldmine.

According to Einstein's theory, the stronger a gravitational field is, the slower time will run in it. Since the orbit of PSR

1913+16 around its companion is elliptical, the two stars are closer together at some times than at others, and the gravitational field alternately strengthens at periastron and weakens at apastron. When the stars are closer together, near periastron, the gravitational field is stronger, so, according to general relativity, the passage of time should slow and, therefore, the interval between pulses should lengthen. Likewise, the pulsar clock ought to regain time when it's traveling in the weakest part of the field around apastron. Taylor and Hulse's measurements completely vindicated Einstein's predictions on this score.[36]

They also explained an effect that would have had Newton completely baffled. Recall the anomalous movement, or advance, of Mercury's perihelion–the extra forty-three arc-seconds per century by which it drifts around the sun above and beyond what Newtonian theory expects. In the strange realm of the Hulse-Taylor pulsar, space-time is vastly more warped, with the result that the periastron advance of the component stars makes Mercury's aberrant motion look like a trivial accounting error. PSR 1913+16's point of closest approach to its partner shifts by an astounding 4.2 *degrees per year*. In other words, its periastron advances in a single day by the same amount as Mercury's perihelion advances in a century. Remarkable though this figure seems, it's exactly in line with the bookkeeping of general relativity.

These data from the Hulse-Taylor system, showing the slowing of time and the periastron drift, merely confirmed other observations made previously by astronomers and physicists that supported Einstein's view on gravity. But another effect was seen for the first time thanks to careful studies of the pulsar's metronomic ticking. In 1983 Taylor and his colleagues reported that there had been a systematic shift in the observed time of periastron relative to that expected if the average orbital separation between the two stars had remained constant. Data taken in the first decade after the discovery showed a decrease in the orbital period of about 76 millionths of a second per year. By 1982, the pulsar was arriving at its periastron more than a second earlier than would have been expected if the orbit had remained constant since 1974. Put simply, its orbit is decaying. For every circuit that PSR 1913+16 completes around its companion, its orbit shrinks by just over a tenth of an inch. And Taylor and Hulse knew exactly why.

Take two big masses that are close together, whirl them around their common center of gravity at high speed, and general relativity is clear about what will happen. The orbital energy of the masses will gradually radiate away. They will give off gravitational waves and, as they do so, their orbits will get smaller and smaller. More than twenty years of watching PSR 1913+16 have shown its orbital period to be dropping at just the rate expected due to loss of orbital energy by gravitational radiation. While that isn't proof of the existence of gravitational waves, it's a very persuasive sign. It was evidence enough, at any rate, to persuade the Nobel committee that Taylor and Hulse should receive the 1993 prize for physics.

In time, the pulsar and its mate will come so close together that they may tear each other apart in their ferocious gravitational embrace. Barring that, a collision is inevitable, some 240 million years from now. A final, powerful burst of gravitational waves will mark the event and, if our descendants are still around and listening, be recorded as a sharp spike in detectors here on Earth.

Gravity's Rainbow

Since Taylor and Hulse's 1974 discovery, several more binary neutron star systems have been found. Many scientists now believe it is from this type of object that the first gravitational waves will be detected directly by a device such as LIGO. We might even be lucky enough to witness the death throes of such a system. In the last few minutes before collision, the stars will spin around each other faster and faster, the frequency of gravitational waves they emit will rise, and the strength of these waves will increase dramatically.

Even more powerful would be the gravitational waves from a binary in which a black hole orbited a neutron star or two black holes circled around each other. Observing them might confirm a curious fact that has emerged from computer simulations. These simulations suggest that no matter how a black hole forms or is disturbed by infalling material, collision with another object, or orbital interactions with a companion, it will "ring" with a unique frequency–its natural mode of vibration. Detecting this unique signature in the form of gravitational waves would not only tell scientists that their theories were on the right track, but it would also supply valuable information about the black hole's size and how fast it is spinning.

Black holes of the supermassive variety, thought to exist at the heart of many galaxies, are expected to leave another characteristic fingerprint of gravitational waves as they dine on the occasional star that comes too close to their lair. Finding such waves would supply direct proof that these dark dwellers of the galactic deeps are real. For physicists, all such studies of gravitation in highly relativistic settings are important because they afford a way to test general relativity in extreme situations (the so-called strong-field limit) where Einstein's theory isn't merely a small correction to Newtonian gravity. Watching for gravitational waves from regions where the curvature of spacetime is pronounced opens up the possibility of fundamentally new science.

Many researchers relish the prospect of what gravitational wave astronomy may do for our understanding of supernovae.

A massive star faces a violent death when it runs out of nuclear fuel. The center of the star can then no longer support itself against gravity and collapses under its own weight. At the same time, large parts of the star explode into space. After the explosion, a neutron star, or in the case of a very massive star, a black hole, remains. A fine scientific reward will follow the detection of gravitational waves from supernovae. We will then be able to understand the final stages of a star's evolution and maybe actually watch neutron stars and black holes being born.

One supernova or so per century is a typical galaxy's standard allotment. Since we're likely to have a long wait between these stellar eruptions in our own galaxy, we'll need to build bigger and more sensitive detectors that will let us see further out into the universe.

There is even the possibility that, with future technology, we'll be able to pick up the faint gravitational waves left over from the Big Bang. That might also tell us something about the future of the universe. In contrast to electromagnetic radiation, it isn't fully understood what difference the presence of gravitational radiation has on a cosmological scale. It may be that a strong enough sea of primordial gravitational radiation, with an energy density exceeding that of the Big Bang electromagnetic radiation by a few orders of magnitude, would have an effect on the rate at which the universe is expanding.

10

Disturbing News

I canna change the laws of physics, Captain—
but I can find ye a loophole.

—MONTGOMERY SCOTT, CHIEF ENGINEER,
STARSHIP *ENTERPRISE*

The humble pendulum seems an unlikely source of mystery. According to some experiments carried out over the past few decades, however, swinging pendulums behave in odd ways from time to time. Meanwhile, billions of miles away, at the edge of the solar system, more strange things are going on. Two of our spacecraft are off course, as if being lured by an unknown and invisible siren. If all other explanations for their errant behavior fail–and some have already done so–these wayward pendulums and probes could spell big trouble for our understanding of gravity.

Wild Swings

The problem with pendulums was first brought to light by the Frenchman Maurice Allais, winner of the 1988 Nobel Prize–not for physics but for economics. In 1954, while at the School of Mining in Paris, Allais decided to do a series of experiments involving a special kind of pendulum over a thirty-day period to investigate a possible link between magnetism and gravitation. What he found was much stranger and more unexpected.

Allais used a Foucault pendulum, which in its original guise is just a big weight on the end of a long wire suspended from a rotating hinge. Named after Jean Foucault, who also invented the gyroscope, this type of pendulum provides a remarkable demonstration of something that happens all the time under our feet. The plane of its swing seems to change, slowly and continuously, so that it completes one whole rotation in space every twenty-four hours. In fact, this motion is an illusion. The truth of the matter is that the plane of the swing remains constant in space while the entire Earth spins beneath it.

Foucault's apparatus, set up at the 1851 World's Fair, was a stunning public tour de force and is considered the first nonastronomical proof of our planet's rotation. Today, with hinges raised to heights of over 90 feet, Foucault pendulums form massive display pieces in the lobbies of more than sixty museums and entrance halls around the world, including the United Nations Building in New York, the Smithsonian Museum in Washington, St. Isaac's Cathedral in Leningrad, and the Pantheon in Paris. Two scientists who eventually went on to win Nobel prizes made Foucault pendulums, at one time or another, a subject of their research: the Dutchman Heike Kamerlingh-Onnes, who took the 1913 physics prize for his investigations of matter at low temperatures (which led to the production of liquid helium), and Allais, who won his Nobel medal for contributions to the theory of markets and efficient utilization of resources.

Allais used what's called a paraconical pendulum–a short, rigid version of the more familiar Foucault design. His month-long study in the summer of 1954 was a marathon of patience and endurance. Every fourteen minutes at his Paris laboratory, deep underground, the pendulum was released and the direction of rotation (in degrees) measured, without ever missing a datum point. By a stroke of good fortune, part of the experiment took place during a total solar eclipse. On June 30 the moon began to pass in front of the sun, its shadow racing across three continents. As the edge of the shadow reached northern France, a very peculiar thing happened in Allais' lab. At the onset of the eclipse, the pendulum's direction of swing suddenly started to rotate backward–the swing plane lurching by a colossal 13.5 degrees. Normally, a Foucault pendulum shifts its angular plane by a modest 0.9 degrees per minute. For the entire duration of the eclipse, however, Allais' pendulum remained more than 13 degrees out of kilter. It was as if the pendulum had somehow been influenced by the alignment of Earth, the moon, and the sun. But how, wondered Allais, could that possibly be?

Five years later, on October 22, 1959, Allais repeated his experiment during a partial solar eclipse–and again watched as the pendulum swung wildly. This time, the results caught the attention of the great rocket pioneer Wernher von Braun, who urged that Allais publish his findings in English. Allais did so, in the journal *Aero-Space Engineering*,[1] noting "anomalies in the movement of the . . . pendulum" and "a remarkable disturbance" during the time when Earth, the moon, and the sun lay in a line. Allais commented that the effect "cannot be considered as due to the disturbances of an aleatory order [that is, chance]. Neither can it be considered as produced by an indirect influence of known factors (temperature, pressure, magnetism, etc.)." If the phenomenon was real, its cause was much more out of the ordinary.

Perhaps not surprisingly, the Allais effect, as it has become known, was either ignored by the scientific community–as troublesome anomalies often are–or put down to experimental error. But Allais was a meticulous researcher; his experiments were well thought out, and he repeated his measurements during two solar eclipses. What's more, his findings have been supported by others, including, in 1961, three researchers in Romania who were completely unaware at the time of Allais' discovery.[16]

In 1970 Erwin Saxl, who founded the company Tensitron (a leading manufacturer of tension meters), and Mildred Allen of Mount Holyoke College, in South Hadley, Massachusetts, studied the behavior of a pendulum before, during, and after a total eclipse. Like Allais, they noted large irregularities at the onset of the eclipse. They concluded in a subsequent paper that "gravitational theory needs to be modified."[33]

Saxl and Allen used a completely different setup than did Allais. Instead of a paraconical pendulum, they measured changes in the period of a torsion pendulum, a massive disk suspended from a wire attached to its center. Rotating the disk slightly caused the wire to twist. When it was released, the disk continued to twirl, first clockwise, then counterclockwise, with a fixed period. But during the eclipse, something different happened: Saxl and Allen's twisting pendulum sped up. In the midst of another solar eclipse in India in 1995, D. Mishra and M. Rao of the National Geophysical Research Institute in Hyderabad observed a slight but sudden drop in the strength of gravity as measured by an extremely accurate gravimeter (in essence, a very accurate set of scales).[22]

Three different types of instruments–the Foucault pendulum, the torsion pendulum, and the gravimeter–each reacted in a puzzling way at the time of an Earth-moon-sun alignment. Each of these instruments effectively measured the acceleration due to gravity at Earth's surface. The Allais effect is a small, hypothetical additional acceleration, so tiny that it would take an apple about a day to fall from a tree branch if it were the only gravitational effect around. Still, if it exists, it's a bit odd that it doesn't show up in other, much more obvious ways. If there were some unexplained aspect of gravity at work, you would expect it to be apparent in the motion of planets and other astronomical objects over long periods of time. And why should the effect make itself felt only during a solar eclipse? The sun, the moon, and Earth are pretty well lined up about once a month at the time of a new moon. If something were happening to gravity on so regular a basis, it seems we would have been amply aware of it by now.

Allais himself never doubted the implications of his results. In his acceptance speech for the 1988 Nobel Prize in economics, he summed up his experiments and noted, as on other occasions, that the phenomenon he observed was "quite inexplicable within the framework of currently accepted theories" of gravity. There is a hitherto unperceived flaw, he thinks, in the general theory of relativity. And what an irony as well as a bombshell that would be since it was observations taken during a solar eclipse that established general relativity in the first place.

There have also been some null results to which physicists will often point when pressed on the issue of the Allais effect. Experiments carried out by Finnish geophysicists, for example, on July 22, 1990, when the eclipsed sun rose above Helsinki, yielded no anomalous readings. However, a thorough analysis of all the evidence gathered over the past half century, carried out by Chris Duif, a researcher at the Delft University of Technology in the Netherlands, suggests that the effect *is* real and unexplained.[6] Duif looked at various conventional explanations for the Allais effect and decided that the most obvious possibility–that it was a mere measuring error–was unlikely because similar results had been found by several different groups, operating independently and, in at least one case, without prior knowledge of Allais' original findings.

Duif also cast a skeptical eye over several explanations that relied on conventional physical changes that might take place during an eclipse. One of these was that the anomaly is caused by the minor seismic disturbance induced as crowds of sightseers move into and out of a place where an eclipse is visible. This is improbable, argued Duif, given that one of the experiments with a positive result was conducted in a remote area of China while another that drew a blank took place in Belgium, one of the most crowded parts of the planet. Duif also considered the possibility that because the moon's shadow cools the air during an eclipse, this cooler, and thus denser, air might exert a different gravitational pull on the instruments. Conceivably, he surmised, this change might affect a gravimeter, but there is no way it could account for the results from the pendulums.

Duif ruled out a third explanation, too: the contention that cooling of Earth's crust due to the eclipse shadow causes the ground to tilt slightly and thus distorts the results. He noted that although a detectable tilt *is* caused when the temperature drops by a few degrees, that tilt is too small to explain the anomalies and, in any case, it would lag roughly half an hour behind the shadow (because it takes time for the ground to cool), while the experimental measurements show an instantaneous change in the acceleration due to gravity during an eclipse.

Although Duif discounted each of the conventional explanations on its own, he admitted they might, in combination, account for the Allais effect. More likely, though, in his opinion, there is something wrong with our understanding of gravity. This suggestion would fit in with another outstanding puzzle: the fact that the Pioneer 10 and 11 space probes, launched by NASA more than four decades ago, are receding from the sun slightly more slowly than they should be.

An Unexpected Detour

March 2, 1972. A sleek white obelisk points at the stars; atop it, a 570-pound bundle of equipment destined to go where no one and no human-made machine has gone before. Dusk falls over the Floridian cape, followed by a clear dark evening. At 8:45 P.M. EST, bright flame and billowing smoke erupt from the tail of the metal needle–an Atlas Centaur SLV-3–and, moments later, the three-stage rocket begins to lift from its pad toward the blackness of space. Early the next day, Pioneer 10 receives its final boost from the remains of the launch vehicle, accelerating it to a record-breaking speed of 32,000 miles per hour. Now it heads out alone, away from Earth, away from the sun, bound for the icy depths of the solar system and the greater universe beyond.

No craft had ever before ventured into the asteroid belt, that rock-strewn region between the orbits of Mars and Jupiter, and there was concern that the little probe might be destroyed by slamming into unseen minor debris that might lie among the known hurtling mountains. Those fears were allayed, however, and the belt region proved navigable for future travelers, when Pioneer passed safely through to the farther edge in December 1972. One year later, it flew past mighty Jupiter, capturing the first close-up images of that giant world during an encounter that took it to within 125,000 miles of the planet. The following year, on December 2, 1974, Jupiter was visited again when Pioneer 11, Pioneer 10's sister craft, swept by. But this time, mission controllers on Earth took the opportunity to target their robot emissary's approach so that it would benefit from Jupiter's gravitational pull to swing it on to a new course that would take it past Saturn almost five years later.

Their planetary flybys complete, the twin Pioneers became the first terrestrial spacecraft to head out of the solar system forever. Not even if every ounce of their remaining fuel had been exhausted in an effort to alter their courses would it have made any difference. Following their triumphant meetings with the two greatest planets in the sun's domain, these two little vessels were set irrevocably on exit trajectories that would lead them, in time, to the immense void between the stars. They were bound for interstellar space.

Pioneer 10 is heading in the opposite direction to that in which the sun moves through our galaxy and will eventually skirt the environs of Aldebaran, a bright orange star that lies sixty-eight light-years away in the constellation of Taurus the Bull. But our descendants will need long memories if they're to celebrate the event. At its present speed of 7.5 miles per second, this most primitive of starships won't enter Aldebaran space for another 2 million years. Its twin, heading away from the sun in virtually the opposite direction, has similarly not much to look forward to over the next few hundred millennia: a relatively near pass of one of the stars in Aquila is on its itinerary–but not until the year 4 million A.D.

Alas, both Pioneers are already dead, their radioisotope power sources having degraded to the extent that they can no longer run any of the onboard instruments or raise a signal home. Mission controllers last heard from Pioneer 11 on November 30, 1995. Pioneer 10 lasted longer. Although its scientific mission officially ended in 1997, telemetry data continued to be received at regular intervals until April 2002, when the probe was about 7.5 million miles away–more than eighty times the distance from Earth to the sun. The last faint signal was picked up by NASA's Deep Space Network on January 23, 2003. Since then, nothing.

Nothing, that is, except a mystery. Long before the Pioneers fell silent, they had been sending back evidence that some unknown hand was at work beyond the orbit of Pluto, nudging the little craft off course. Not that it was obvious. Only in 1998 did mission scientists and engineers at NASA's Jet Propulsion Laboratory (JPL) in Pasadena, California, finally recognize that both Pioneers were not quite where they should have been. Both were ever so slightly veering off their expected courses and slowing down just a tad more than predicted, as if the solar system were tugging marginally too hard on the two vehicles. The unexplained deceleration was tiny: less than a nanometer (a billionth of a meter) per second per second. That's equivalent to just one ten-billionth of the gravity at Earth's surface, but it is unrelenting, and its effects have mounted up over time. When mission controllers last heard from Pioneer 10, it was a quarter of a million miles from where it was supposed to be–roughly the distance from Earth to the moon. When NASA lost touch with Pioneer 11, several years earlier, it was heading for a similar deviation.[3]

A raft of mundane explanations need to be considered for what's been going on. Perhaps thermal radiation from the onboard power sources or minor fuel leaks were supplying a bit of extra thrust that would account for the anomalous motions. Or maybe there was a fault with the spacecraft control systems–a software error, for example. Painstakingly, JPL mission staff analyzed every possible cause by which the spacecraft might systematically be affecting their own trajectories; one by one, they eliminated them. They reexamined other known influences on the probes, including solar radiation. No conventional account of the "Pioneer anomaly" seemed to work. And so, researchers began to look beyond the conventional, at the unusual, and even the bizarre.

Might there be, wondered Chris Duif, some connection between this new anomaly and the Allais effect? The reason von Braun had become so interested in the Allais effect, back in the 1950s, was his suspicion that the strange pendulum results might shed light on some puzzling deviations in the paths of spacecraft at that time. Those awry trajectories were eventually accounted for by more accurate calculations based on known variables. But the Pioneer anomaly has proved more resilient. And, noted Duif, the feeble but persistent force felt by both Pioneer probes is about the same size as that measured by some gravimeters during solar eclipses.

Theories and Speculations

One exotic possibility that has been put forward to explain the Allais effect is known as Majorana shielding. Named after the Italian physicist Quirino Majorana, who worked on the concept in the 1920s, it postulates that large masses (such as the moon) can partially block the gravitational force from more distant objects (such as the sun).

The origins of this idea of gravitational shielding go way back to Nicolas Fatio de Duillier, the Swiss mathematician and one-time close friend of Newton. When Newton admitted he didn't know how gravity worked, de Duillier suggested, in 1690, that it arose as a shadowing effect associated with the absorption by material bodies of minute particles. This "push" theory of gravity was then developed further, in the eighteenth century, by another Swiss mathematician, George-Louis

LeSage, also remembered for building and patenting the first electric telegraph. LeSage believed there was some kind of pressure in space. Masses, he thought, shielded one another from this space pressure and were thus pushed together by the unshielded pressure on their opposite sides. Although LeSage's theory never won much support in the wider scientific community, it did strongly influence John Herapath, an English amateur scientist, in developing an early version of the kinetic theory of gases. It also came back into play when attempts were made to explain some anomalies in the motion of the moon that had been detected in the first decade of the twentieth century by Simon Newcomb.

In 1912 the German astronomer Kurt Bottlinger calculated the effects that would occur if the gravitational force between the sun and the moon decreased during lunar eclipses. What he found was a fluctuation in the moon's longitude that agreed with Newcomb's observations. Subsequently, Bottlinger's results were criticized by the Dutch astronomer Willem de Sitter, and Einstein tried to supply an alternative explanation in terms of changes of the Earth's rotation due to tidal effects. Einstein's analysis was soon proved to be wrong, however, and the moon's anomalous movements went unexplained for many years.

In the 1930s the mystery disappeared from view when astronomers began to use so-called ephemeris time, which was defined in a way that assumed the motion of the moon to be regular. Even before this, the widespread acceptance of general relativity undermined belief that an effect involving gravitational absorption could exist, pulling the rug out from any further experimental and astronomical studies of this hypothesis. But it didn't stop Majorana. In the 1920s, the decade in which general relativity came of age, he did a series of lab experiments involving lead and mercury shields, in which he reported a small gravitational absorption effect.

There's a need for Majorana's research to be repeated to check if the shielding he found is real. If it is, it may be relevant in the context of the Allais effect. But it's hard to see how it would have any bearing on the Pioneer observations.

Another maverick theory of gravity, called Modified Newtonian Dynamics (MOND), is the 1983 invention of Mordehai Milgrom of the Weizmann Institute of Science in Rehovot, Israel.[21] In MOND, Newton's inverse square law applies only where gravity is strong; where it's weak, there's a more gradual drop-off with distance. Some physicists have been drawn to this idea because it might explain the anomalous rotation of galaxies (without the need for the more dramatic thesis we'll run into in the next chapter) and even point the way to a successful merger of general relativity with quantum mechanics. Its biggest problem is that it contradicts the well-tested predictions of relativity, and so it doesn't work well in situations where objects are moving close to the speed of light or under high accelerations. Because of this, MOND hasn't been able to say anything useful about, for example, black holes or how the universe came into being in the Big Bang. To try to remedy that defect, Jacob Bekenstein of the Hebrew University in Jerusalem has found a way around the problem.[4] While the familiar, conventional force of gravity gets weaker in proportion to the square of distance, Milgrom suggested that there is also another form of gravity whose strength diminishes more slowly, declining linearly with distance.

In MOND, a key parameter is the acceleration, a_0, below which gravity supposedly switches from its weaker to its stronger form. To fix MOND's clash with general relativity, Bekenstein brought in twin fields, one giving rise to conventional gravitational phenomena, the second serving as the arena for phenomena involving the other fundamental forces of nature—electromagnetism and the strong and weak nuclear forces.

Bekenstein claimed his theory is consistent with general relativity, Newtonian gravity, and MOND. It reduces, he said, to Einstein's theory at high speeds and accelerations well above a_0, to Newtonian gravity at low speeds and accelerations above a_0, and to MOND at accelerations below a_0. Interestingly, it predicts some subtle new effects that might modify gravity in the outer solar system, although whether it will have any bearing on the Pioneer anomaly, it is too early to judge.

New Probes

To solve the riddle of the wayward Pioneers and see if this is connected somehow with the misbehaving pendulums reported by Allais and others, we need more data. How does the Pioneer anomaly, if it's real, vary with distance and direction from the sun? Over what range does it apply? One way to begin tackling these problems is by observing other objects that travel through the same region of the outer solar system as the Pioneers. These include far-flung asteroids, some of which are large and bright enough to be tracked from Earth for at least part of their journeys around the sun.

Gary Page of George Mason University in Fairfax, Virginia, and his colleagues have identified fifteen asteroids that might also be subjected to the same mysterious force as Pioneers 10 and 11.[28] All have orbits that stretch into the "anomaly zone"—the region beyond about twice the distance from the sun as Saturn at which deviant motions in the spacecraft began to be noticed. Of the fifteen candidates identified by Page and his coworkers, the best is 1995SN55. This 230-mile-wide space rock has spent the past half-century in the anomaly zone, and so it should have experienced the biggest perturbation. Tantalizingly, 1995SN55 isn't where predictions say it should be. We can rule out fuel leaks and other artificial factors in the case of big, dumb objects like asteroids. So, whatever is causing 1995SN55 to stray off course is natural. Only more work will tell if it is being subjected to the same influence as the Pioneers and whether it might be possible to use such observations to discriminate among the various proposed explanations of the Pioneer anomaly. Even if there's no connection with the spacecraft puzzle, watching remote asteroids closely will supply valuable information at ranges over which there has been a dearth of test data for use in gravitational studies.

To get to the bottom of the mystery, though, we may end up having to send out more spacecraft. An outline proposal has already been made for an experiment specifically designed to test unusual gravitational effects in the outer solar system that could ride piggyback on a future deep space probe.[2] Perhaps it will find nothing awry in the cold, lonely depths beyond Pluto. Perhaps it will find that gravity there behaves subtly different than we'd thought. Or, perhaps it will find that there's more to the darkness of space than meets the eye.

11

In the Dark

Science has "explained" nothing; the more we know, the more fantastic the world becomes and the profounder the surrounding darkness.

—ALDOUS HUXLEY

Astronomers are seriously embarrassed; the universe is playing hide-and-seek with them—and winning. Imagine if geographers were forced to admit that, contrary to previous claims, nine-tenths of the earth had been overlooked and some new continents would need to be added to the map books (just as soon as they had been tracked down). Or that anatomists had to break the news that, after centuries of effort and millions of dissections, they had found evidence for an enormous organ in the body that had somehow previously escaped their attention. About thirty years ago, astronomers began to realize they were faced with owning up to exactly this kind of oversight—the universe they could see was only the tip of an enormous cosmic iceberg.

There was, it turned out, far more matter out there than anyone had realized. Its gravitational effect dwarfed that of all the luminous material, in the form of stars and bright gas, that we could see. But what this massive hidden component consisted of was anybody's guess.

Now it's happened again. Just as we were getting used to the fact that a big chunk of everything was playing hooky, astronomers have had to report another serious lapse in accounting. Not only is most of the matter in existence hidden from view, but there is an awesomely powerful force at work opposing gravity. This force may be crucial to the fate of the universe, and we have only just noticed it. It's a bit worrying. We thought we lived in a cosmic house that had just one big room–a room containing ordinary matter. Now we find it has two other rooms, and we're missing the keys to both of them.

A Matter of Concern

The whole idea that there's a lot more to the universe than meets the eye has its roots in some observations made by the Caltech astronomer Fritz Zwicky in the 1930s. Zwicky, Bulgarian-born but of lifelong Swiss nationality, was one of the most outlandish scientific characters of the twentieth century. He studied at the ETH in Zurich while Einstein was teaching there, and then came to the United States in 1922 at the invitation of Caltech's Arthur Millikan. Anecdotes about Zwicky, mostly to do with his outrageous behavior and intolerable rudeness, would fill a chapter. His popularity wasn't enhanced by his habit of calling other astronomers at the Mount Wilson Observatory, where he worked, "spherical bastards." Why spherical? "Because they were bastards," said Zwicky, "when looked at from every side." His first words upon bumping into a new student he didn't know were "Who the devil are you?" It's perhaps appropriate that this bizarre but influential figure should have made a reputation by studying unusual galaxies.

In 1933 Zwicky stumbled across a puzzle. He'd been looking closely at galaxies in the Coma Cluster, a big gathering of star cities, several thousand strong, that sits about 290 million light-years away in the direction of the constellation Coma Berenices. From measurements of how the galaxies were moving near the edge of the cluster, he came up with an estimate for the cluster's total mass. But when he compared this value with one based on the number of galaxies and the total brightness of the cluster, he found that it was too high *by a factor of 400.* The gravitational tug of the visible matter in the cluster wasn't strong enough–not remotely strong enough–to explain the high speed of galaxies at the periphery. "[W]e arrive at the astonishing conclusion," said Zwicky, "that dark matter is present with a much greater density than luminous matter." Three years later, Sinclair Smith's observations of the Virgo Cluster revealed another huge mass discrepancy similar to that found by Zwicky.

Dark matter had been discovered–and named. Yet no one else at the time paid it the slightest attention. Part of the reason was that scientists knew ways that observations such as those of Zwicky and Smith could be misleading. Maybe a galaxy with a particularly high velocity isn't gravitationally bound to the cluster at all; it could be on its way out, having exceeded the escape velocity, or perhaps it had never been part of the cluster but was just sailing through. Maybe some of the observed high-speed galaxies lay in the foreground and just seemed to be part of the cluster because they fell along the same line of sight. Doubts like these kept the whole business of Zwicky's dark matter off the scientific radar.

Although no one was yet putting two and two together, evidence quietly continued to build up. In 1939 Horace Babcock looked at the rotation of stars in the Andromeda Galaxy, the nearest big galaxy to our own. A lot of nonglowing matter, he concluded, had to be present in the outer parts to explain the curiously fast stellar motions there. Jan Oort found the same kind of result in 1940 with another galaxy. In 1959 Franz Kahn and Lo Woltjer looked at movements within the Local Group, the little galaxy cluster to which the Milky Way and the Andromeda spiral belong and showed that most of the mass in the Local Group was invisible. Perhaps, they speculated, it took the form of very hot gas.

As the 1970s dawned, several researchers, including Vera Rubin, an astronomer working at the Carnegie Institution's department of terrestrial magnetism in Washington D.C., looked in more detail at how stars rotated in Andromeda. By the middle of the decade the conclusion that large amounts of unseen matter lurk in the outskirts of galaxies was being supported by theoretical work on galactic structure and stability. At last, the connection with the earlier work of Zwicky and Smith was made, and the problem of dark matter burst onto the scientific stage. Most of the mass of galaxies, and therefore of the universe, had apparently been overlooked. Shaking their heads in disbelief, cosmologists beat a hasty retreat to the drawing board.

Two Puzzles and a Solution

The future of the universe used to be a fairly straightforward problem. Just tot up all the luminous matter throughout space (mostly in the form of stars), and if there were more than a certain amount, the universe would eventually reverse its outward rush and collapse back to a Big Crunch, many billions of years from now. Alternatively, if matter was in too short supply, gravity would lack the strength to prevent the expansion going on forever. The key factor was Omega. Depending on whether Omega was less than one, equal to one, or greater than one, the universe would correspond to the open, flat, or closed space-time scenarios first described by Friedmann in the 1920s. Only by adding up the total amount of matter in existence or–what amounts to the same thing–working out the average density of cosmic matter could we decide which kind of space-time we live in. But now that calculation seemed to have been complicated by the discovery of dark matter–another unknown to be factored in.

On a seemingly different front, cosmologists in the 1970s were struggling to explain certain overall properties of the universe today in terms of Big Bang theory. Chief among their concerns was that the universe is a lot "flatter" and smoother than ideas about the Big Bang suggested it should be. The so-called flatness problem stemmed from the fact that although the value of Omega wasn't known exactly, it was certainly known to be in the ballpark of unity. For the universe to be even roughly this flat now, Omega would, according to Big Bang theory, have to have been well within *one part in a trillion* of unity at the dawn of time. This looked like an almost unbelievable case of fine-tuning. If Omega hadn't been within a hair's-breadth of unity at the cosmic kickoff, space-time would either have collapsed in on itself long before now or would have rushed out so quickly that there wouldn't have been time for galaxies and other collections of matter to form–and we wouldn't be here to tell the tale.

The other big conundrum, called the smoothness problem, was why the universe looks pretty much the same in every direction; overall the cosmos is surprisingly uniform in appearance. To explain that kind of uniformity today, cosmologists had to assume that the primeval soup was virtually devoid of any kind of lumpiness. At the same time, it couldn't have been perfectly smooth; otherwise, there wouldn't have been any seeds from which future galaxies could grow. Again, it seemed as if the universe was balanced on a knife-edge without any explanation of why that should be so.

Then, beginning in 1979, a remarkable solution to the flatness and smoothness problems began to emerge. Alan Guth, a physicist at MIT, and others suggested that for a very brief period just after the Big Bang, the universe went through a sudden and astonishingly rapid growth spurt called inflation. This blew up a tiny region of space-time into the universe we inhabit today, producing, as a natural consequence, the kind of flatness and smoothness we see around us. Inflation took away the need to believe in fantastic coincidences. But one of the essential byproducts of this update to Big Bang theory was a prediction for the value of Omega. Inflation insists that Omega is equal to 1–not just close to 1, but exactly unity.

That seemed to put a question mark over whether inflation could be true. Astronomers knew there wasn't enough ordinary (bright) matter around to push Omega anywhere near close to 1; the shortfall was huge. With the discovery of dark matter, however, researchers began to ask if this didn't hold the key to the Omega riddle. Could a *combination* of dark and light matter give the unity value for Omega that inflation demanded? Alongside this question was another, equally compelling question: what on earth *was* dark matter?

The Stuff of Dreams

We already know about some dark matter. Planets (small, such as Earth, or big, such as Jupiter), moons, asteroids, and other minor debris, plus all the nonluminous gas and dust between the stars come into this category of ordinary dark matter. So, too, do all the very lightweight stars and so-called brown dwarfs (stars that don't have enough mass to make energy by fusing hydrogen into helium) that make up a lot of the stellar population of a galaxy like the Milky Way but contribute almost nothing to its brightness.

Part of all the dark matter that exists is bound to consist of familiar objects and materials such as these. But there is very little chance it could be a significant part. The reason for this has to do with what happened in the Big Bang some 13.7 billion years ago. At the birth of the universe, all that existed was just an extremely hot soup of all sorts of particles. As the universe grew and cooled, the ordinary matter particles such as neutrons, protons, and electrons started to join together to form atoms of the elements we see around the cosmos today—predominantly hydrogen and helium. Our theory of element making in the first few minutes of cosmic genesis, called Big Bang nucleosynthesis (BBN), has been a great success. Not only does it predict that hydrogen and helium should be dominant but it gets them in the right proportions we see, allowing for changes that have happened inside stars since.

But with this success comes a catch: the amount of each element that forms, it turns out, depends very sensitively on the amount of baryonic matter that the universe had available. Baryons are a family of heavyweight particles that includes the proton and the neutron. Basically, BBN predicts the right ratios for the elements in the universe today only if the original amount of baryonic matter was no more than 10 percent of the critical amount of matter needed to stop the cosmic expansion. Because scientists believe dark matter makes up far more than 10 percent of the critical value, they're pretty confident that most dark matter isn't made of baryons. Of course, it could be that, despite having done such a good job so far, BBN will turn out to be flawed; some researchers are looking into that possibility. But the prevailing view is that whatever dark matter is, it isn't made of the same stuff we are.

So we have to look at the alternative. If dark matter isn't

made of ordinary matter, which is baryonic, then it must consist of some kind of "exotic matter," which is nonbaryonic. Theorists use that term *exotic* as a catchall; it might mean something strange and new, but it doesn't have to. Some nonbaryonic dark matter might consist of a particle that has been known for many years: the neutrino. Wraithlike, billions of neutrinos from space (many from the sun) pass through your body, and then, every second, clean through the entire Earth. Although aloof, they're one of the most populous particles in the universe, and in recent years it has been found that, contrary to previous belief, they almost certainly have a small amount of mass.

For a long time, physicists thought that neutrinos, like photons, traveled at the speed of light and therefore must have zero rest mass. But the discovery that neutrinos can change from one form to another, made within the last few years, implies that they can't be massless. Their mass, if confirmed, is certain to be tiny. But even if it's as small as one five-thousandth the mass of the electron (which, in turn, is almost 2,000 times lighter than a proton) then, given the vast number of neutrinos cosmoswide, it would still amount to enough dark matter to reverse the cosmic expansion. As it happens, there's a good reason, as we'll see in a moment, to suppose that neutrinos don't make up the lion's share of dark matter.

The other big nonbaryonic dark matter hopeful goes by the unprepossessing acronym WIMP (weakly interacting massive particle) and belongs to a class of hypothetical heavy particles that hardly interact at all with familiar forms of matter (otherwise, they would have been discovered by now). You might think that since no one's ever seen an axion, a neutralino, or any of the other WIMP candidates, there would be a lot of skepticism surrounding this dark matter option. On the contrary, in the contest to be gravitational kings of the universe, WIMPs seem to be winning out over their biggest rivals, the MACHOs.

MACHO stands for massive compact halo object (the name chosen in 1991 deliberately to contrast with WIMP). According to the MACHO point of view, sizable galaxies such as our own are cocooned by dark matter halos that have a hefty population of nonluminous objects–the kinds of things we were talking about earlier: brown dwarfs, other types of very dim stars, black holes, planets, and so on. Although we wouldn't be able to see MACHOs directly, we could, astronomers realized, hope to detect them in another way. Any concentration of matter can, under the right circumstances, act as a lens, bending and focusing the light rays from a source that lies behind them at a much greater distance. Since the early 1990s, a number of projects on the lookout for "microlensing" events that would give away the presence of MACHOs have had some success. In 2006, an extrasolar planet only five times as massive as Earth was found by this technique. But overall, not enough MACHOs have been found to account for more than a fifth of all dark matter, and the final figure may be much less. Also, because MACHOs are presumed to be made of ordinary (baryonic) matter, their contribution is likely capped by the BBN restriction mentioned earlier.

Weighing the Possibilities

The case of the missing matter has yet to be solved, but at least we have a good idea who the main suspects are: ordinary nonluminous objects (mostly in the form of MACHOs), neutrinos, and WIMPs. The strange thing is this: although we know for certain that the first two suspects are real and make *some* kind of contribution to the dark matter total, it is seeming more and more inescapable that the one that no one's ever detected–WIMPs–are the biggest component of all.

We've seen that dark matter can be either baryonic or

nonbaryonic and that BBN (assuming it's correct) limits the amount of baryonic matter in the universe. That rules MACHOs out as a major dark matter component. Another way to categorize dark matter is into hot and cold varieties. Very light stuff that moves around at near light-speed is called hot dark matter (HDM), and the top candidate for this is neutrinos. We also know that whatever dark matter is, it is the primary source of gravitation in the universe; therefore, it must have played a crucial role in determining the structure of everything from galaxies to the large-scale arrangement of superclusters (clusters of clusters) of galaxies.

We know there is a lot of dark matter in galaxies because of the way that stars move in them; we also know there is a lot of dark matter in galaxy clusters because of the way galaxies move in *them.* HDM wouldn't stay clumpy in this way, nor would it have clumped together in the first place to play the important part it undoubtedly has in galaxy and galaxy cluster formation. These conclusions have been totally supported, in recent years, by results from the Wilkinson Microwave Anisotropy Probe (WMAP). It now seems beyond doubt that HDM can make up only a small fraction of the dark matter total.

But when it comes to cold dark matter (CDM), the only viable candidate we have that can account for the bulk of CDM are WIMPs. So, we're pretty much driven to the conclusion that WIMPs account for most of the dark matter in the universe. The only worrying thing, as we've mentioned, is that all this is based on inference–we don't have the slightest observational evidence that WIMPs exist.

As theories, both CDM and HDM have their problems. HDM can't form lumpy structures such as galaxies and galaxy clusters, and CDM has problems forming the biggest structures of all, such as the vast walls along which astronomers have found superclusters of galaxies to be arranged. Most likely, the final answer lies with some kind of mixed dark matter (MDM), in which both CDM and HDM are indispensable.

Research into dark matter, on both the theoretical and observational fronts, is barreling along and turning up new surprises along the way. An intriguing prediction of theories about the Big Bang and the large-scale structure of the cosmos is that "dark galaxies"–galaxies made almost entirely of dark matter–should be very common. These starless objects would have formed from pockets of pure dark matter that are expected to have survived from the earliest days of the universe. In 2005 astronomers in the United Kingdom announced the discovery of the first dark galaxy, VIRGOHI21, lying 50 million light years away in the Virgo Cluster. Radio observations of the rotation of hydrogen gas in VIRGOHI21 have revealed there must be about a thousand times as much dark matter as hydrogen in this galaxy and that its total mass is about one-tenth that of the Milky Way.

In February 2006, following its study of twelve nearby dwarf galaxies, the Cambridge University team sprang a surprise. They found evidence that dark matter is neither very hot nor very cold but somewhere in between, with a temperature of about 10,000 degrees Celsius. They also found that, whatever the size of the galaxy, dark matter seems to come in "magic volumes"–standard size packages–about 1,000 light-years across.

Dark matter rules ordinary matter in the cosmic scheme of things. But is there enough dark matter around to make up the shortfall and propel Omega to 1, as inflation theory demands? The answer is a resounding no. Both theory and observation point decisively to the fact that, although there is about six times more dark matter in the universe than ordinary matter, even the two together can't push Omega up to the magic value of unity. Fortunately for members of the Inflation Theory Preservation Society, there is even more to the universe than meets the eye.

A Cosmos out of Control

We've known since the 1920s that the universe is expanding–no doubt about it. No doubt, too, that this leaves only two possibilities for the far future: either the expansion will some day stop or it will carry on forever, albeit, one might suppose, at a slower and slower pace. What no one had reckoned with was that the growth of the universe might actually be *speeding up*. How could it? The only issue had been how much gravity there was in all of space. The universe has a brake pedal–we just didn't know how effective it was. How could the rate of cosmic expansion be increasing? Astonishingly, we've found that the universe also has a gas pedal, and it's being pressed down even harder than the brakes.

It's always comforting to have a name for something, even if we really don't know what we're talking about. The mysterious, invisible stuff that holds galaxies together earned the name dark matter. So perhaps it was inevitable that the mysterious invisible property that's been found to be prizing the universe apart should be called "dark energy," a term coined by the University of Chicago cosmologist Michael Turner.

Dark energy burst onto the astronomical scene in 1998 after two groups of astronomers, at the Lawrence Berkeley National Laboratory in California and the Australian National University in Canberra, surveyed dozens of type Ia supernovae, the most luminous kind, in distant galaxies.[29, 31] They found that the supernovae were dimmer than they should have been, implying that they were also farther away–up to 15 percent more remote–than they should have been. The only way for that to happen, the astronomers realized, was if the expansion of the universe had sped up at some time in the past. At first, other scientists questioned the result. Perhaps, they suggested, the supernovae were dimmer because their light was being blocked by clouds of interstellar dust. Or maybe the supernovae themselves were intrinsically fainter than had been suspected. But with careful checking and more data, those explanations have largely been cast aside, and the dark energy hypothesis has held up.

According to the supernova observations, the cosmological acceleration stepped into gear around 5 billion years ago. Before that, it is thought that the expansion was slowing down due to the combined attractive influence of dark matter and the ordinary luminous kind. Theory suggests that the density of dark matter in an expanding universe falls more quickly than dark energy, until the dark energy dominates. For every doubling in the volume of the universe, the density of dark matter halves, but the density of dark energy stays nearly or exactly the same.

As in the case of dark matter, astronomers are interested in what dark energy means for the future of the universe. Because of Einstein's mass-energy equivalency, we can treat both matter and energy as contributors to the overall cosmic mass-energy balance. And when we do this and take into account the latest results on the cosmic microwave background from the WMAP orbiting observatory, we find something quite extraordinary: it seems there's exactly the right amount of dark energy in the universe, when taken together with dark and ordinary matter, to bring Omega to 1. Broken down by ingredient, the cosmic recipe reads: ordinary matter (mostly in the form of hydrogen and helium), 4 percent; dark matter, 23 percent; and dark energy, 73 percent.

A Reprise for Lambda

Dark energy looks uncannily like Einstein's cosmological constant–the Lambda term that Einstein implanted in his field equations to hold the universe in perfect balance–and later regretted. If the comparison is good, then it means we need to think of dark energy as a basic property of what we normally think of as empty space–a property that causes negative pressure, a tendency to inflate space from within. If dark energy equals the cosmological constant, then, implicitly, it has the same density everywhere in the universe and is changeless with time. Think of it as the cost of having space: with every volume of space comes some intrinsic, fundamental energy.

One of the biggest problems to do with the cosmological constant is that there's no known way to derive its value–thought to be about 10^{-29} grams per cubic centimeter–even roughly from particle physics as we presently understand it. In fact, current attempts to figure out the energy of the vacuum give ridiculous answers up to 120 orders of magnitude too large. This monstrous value would need to be canceled almost, but not exactly, by an equally large term of the opposite sign to give the tiny value for the cosmological constant that fits with observations on the large scale. Lacking a knowledge of how to do this doesn't mean we're on the wrong track with the cosmological constant, but it does mean there are gaping holes in our understanding of how to link large-scale effects with what's happening at the subatomic level.

A completely different idea for dark energy is our old–very old–friend, quintessence. Plato's celestial element has undergone a radical makeover and become a cosmoswide field. Unlike the cosmological constant, however, this field can vary in time and space. Generally, it predicts a slightly slower acceleration of the expansion of the universe than does the cosmological constant. In some versions of the theory, however, the energy density of quintessence can increase with time. The most dramatic situation would involve a form of quintessence known as phantom energy that could result in a

Big Rip—a monstrous tear in the fabric of space-time that would destroy the universe as effectively as slicing a balloon with a knife.

A Choice of Futures

If the cosmic acceleration that astronomers are seeing at present were to continue indefinitely, there wouldn't be a lot to look forward to (over many billions of years, that is). Space would continue to expand, pulling the material contents of the universe further and further apart. Our local supercluster would probably be immune from this inexorable stretching—local gravity overriding the general recession. But other clusters and superclusters of galaxies outside our own would gradually move beyond our horizon, slipping out of sight as their relative speed surpassed that of light. This superluminal recession wouldn't break the rules of special relativity because the effect couldn't be used to send a signal between the receding galaxies. Rather, it would prevent any communication between them, and the objects would simply pass out of contact.

In a forever-expanding cosmos, the Earth, the Milky Way, and the Virgo Supercluster would remain essentially undisturbed—yet more and more isolated. Slowly and inevitably, all the useful energy in our island of matter would be used up, and we would suffer the heat death once foreseen for an infinite-sized universe before anyone knew of cosmic expansion, let alone cosmic acceleration.

But we can't be sure that the acceleration will continue more or less as it is at present. The phantom energy version of quintessence would threaten *divergent* expansion, which might pull apart the surface of space-time, hurling us into instant oblivion. On the other hand, dark energy might dissipate with time or even become attractive. Such uncertainties leave open the possibility that gravity might yet rule the day and lead to a universe that contracts in on itself in a Big Crunch.

Which brings us back to the nature of dark energy. Does the cosmological constant or quintessence provide the better description of it? Some quintessentialists think that the best evidence for their theory may come from violations of Einstein's equivalence principle. Perhaps the pendulum anomalies already noted, or some similar puzzling measurements, will win the day for them.

Then there's the puzzle of why the cosmic acceleration should have kicked in when it did. If the acceleration had started earlier in the universe, structures such as galaxies would never have had time to form and life, at least as we know it, would never have had a chance to exist. Proponents of the anthropic principle, which claims that the universe has to be the way it is because we live in it, view this as support for their arguments. However, many models of quintessence have a so-called tracker behavior, which solves this problem. In these models the quintessence field has a density that closely tracks (but is less than) the radiation density until the Big Bang, which triggers quintessence to start behaving as dark energy, eventually dominating the universe. This naturally sets the low-energy scale of the dark energy.

The consensus, though, is behind the cosmological constant. Measurements by the orbiting Chandra X-ray Observatory, for example, in May 2004, square with the idea that dark energy doesn't change much over time, as would be expected if the Lambda interpretation were correct. It would be a pleasing conclusion: the factor that Einstein invented to hold his model universe in stasis proving to be the key to understanding the origin and future of the real cosmos.

12

All Together Now

There are grounds for cautious optimism that we may now be near the end of the search for the ultimate laws of nature.

—STEPHEN HAWKING

Science has a tough problem: its two biggest theories each work superbly on their own, but they don't fit together. Quantum mechanics and general relativity each emerged in the first two decades of the twentieth century. Both revolutionized physics in their own way: quantum mechanics at the level of atoms and the even tinier particles inside them, general relativity on the scale of stars, galaxies, and the whole universe. Both have been compellingly supported over the years by countless experiments, but physicists are painfully aware that these two great theories can't be the final word because they're fundamentally mismatched. It isn't that they predict different results; rather, they're like different pieces of equipment that can't be connected together. We lack a mathematical interface and haven't a clue how to build one. Hence the search is on for the holy grail of physics: an all-embracing theory–a theory of everything (TOE)–that will bring gravity into the quantum fold.

The Lure of One

To find unity in the dazzling multiplicity of things, to forge a common framework of understanding, is an ancient quest not just confined to science. It seems to run in our blood. But in physics, more than any other field, that quest has been a driving force behind new discoveries and an overarching goal for the future. Thales of Miletus was early on the unification scene when, in the sixth century B.C., he argued that water lies at the basis of everything. Water irrigates the land and nourishes plants, it quenches human thirst, and it was home to the fish that Thales ate. Of all substances known to him, water took the most forms: liquid, solid, cloud, and steam, not to mention rocks, which the Greeks believed were made of frozen water (the word *crystal* comes from the Greek *krustallos*, meaning "ice").

Anaximenes of Lampsacus, also in the sixth century, took up the theme of unification. For him, though, the primordial essence wasn't water but air. After all, air is what we breathe, what sustains us throughout life. His contemporary, Heraclitus of Ephesus, plumped for the element fire because "all things are exchanged against fire, and fire against all things." But Anaximander, a pupil of Thales, disagreed with these views. He didn't think that any known substance could be the basal stuff of the cosmos. There was no way, he argued, that fire could form from water, or vice versa, because every observation showed the two to be incompatible. So, for him, the cosmic commonality must be something else–an infinite, eternal substance that embraced the world in its entirety. This ethereal substrate Anaximander called *apeiron*, which means, simply, "boundless."

Pythagoras and his slightly crazy band of followers were also early on the TOE trail, insisting that mathematics–numbers, especially–underpins the kaleidoscope of physical phenomena. And Aristotle, too, played his part in the business of unification by formulating principles (albeit flawed) for all motion on Earth.

But the first great cosmic synthesis in the modern sense would have to wait another twenty centuries for Isaac Newton, who built on the work of Kepler and Galileo (not to mention Hooke, Boyle, and others). In Newton's hands the whole Aristotelian concept of movement was shattered. As the historian Richard Westfall has pointed out, "To Aristotle, to move was to be moved. The motion of any body required a moving agent." Newton brought to center stage the notion of inertia, which allowed motion without cause. Galileo had already laid siege to Aristotle's distinction between "natural" and "violent" motion; Newton completed the demolition. And just as he unified all types of terrestrial movement, so he showed that there aren't different rules for Earth and what lies beyond it; the law of gravitation is truly democratic.

In the second half of the nineteenth century the marriage of electricity and magnetism, officiated by Maxwell, took place. Subsequently, the Scotsman brought together electromagnetism and optics by showing that light is just a form of electromagnetic radiation. Never before had so many phenomena owed so much to so few laws, summarized in just four relatively simple equations. And if electricity and magnetism–two seemingly disparate forces–could be amalgamated, then why not also gravity?

In 1919 Einstein received a curious letter from the German mathematician Theodor Kaluza. In it Kaluza pointed out that if general relativity is extended to a *five*-dimensional space-time, an extraordinary thing happens: the equations can be separated out into ordinary four-dimensional gravitation plus an extra set, which is equivalent to Maxwell's equations for the electromagnetic field. Electromagnetism is then explained as a manifestation of curvature in a fourth dimension of physical space, in the same way that gravitation is explained in Einstein's theory as a manifestation of curvature in the first three. The reason the extra spatial dimension isn't apparent to us, argued Kaluza, is that it's curled up into a fantastically small circle and thereby rendered unobservable. Einstein was impressed and said, "The idea of achieving [a unified theory] by means of a five-dimensional cylinder world never have dawned on me. . . . At first glance I like your idea enormously."

Seven years later, the Swedish physicist Oskar Klein independently came up with an improved version of Kaluza's model. In this version, Klein suggested that the circular nature of the fifth dimension is the origin of charge quantization–the fact that all electric charges appear to be exact multiples of the charge on the electron. It seemed almost too good to be true.

For the last thirty years of his life, spurred on by the higher dimensional visions of Kaluza and Klein, Einstein struggled to combine electromagnetism and general relativity into what he called a unified field theory. His only reward for this lengthy, quixotic venture was disappointment; his effort ended in failure and his sad isolation from the mainstream of physics. "I have become a lonely old chap," he wrote to a friend in the early 1940s, "who is mainly known because he doesn't wear socks and who is exhibited as a curiosity on special occasions." Others were, understandably, far more excited by the possibilities of quantum theory, the central premises of which Einstein utterly rejected, though, ironically, he had been a quantum pioneer and had won his Nobel Prize for this work, not for relativity.

Steps to Unity

As more became known about the goings-on inside the nuclei of atoms and of the way subatomic particles interacted and changed from one form into another, it became clear that there were two other fundamental forces at work in nature besides gravity and electromagnetism. They are known as the strong and weak forces–everyday names for effects that are remarkably well hidden from everyday view. Both are important only over tiny distances, such as those found within atomic nuclei. That there must be another force, more powerful even than electromagnetism, was recognized in 1921 by the Englishman James Chadwick (discoverer of the neutron in 1932) and the Swiss physicist Etienne Bieler. This strong force had to be able to bind together the contents of the nucleus in spite of the determined attempts of the positively charged protons to hurl themselves apart. The Italian-American physicist Enrico Fermi first recognized the weak force in the 1930s; among other things, Fermi realized, it is responsible for radioactive decay.

While Einstein had spurned anything to do with quantum tomfoolery in his efforts to unify gravity and electromagnetism, physicists at large strove to bring the weak and strong forces under the same umbrella as electromagnetism by making full use of the science of the ultrasmall. Quantum physics–quantum *mechanics* as it's generally known (to contrast it with the Newtonian variety)–is all about dealing with energy transactions in the form of minuscule bits called quanta. Scratch beneath the surface and it's a very weird subject indeed, full of counterintuitive ideas that Einstein found completely unbelievable. Yet most scientists managed to turn a blind eye to its more bizarre implications and simply continued to use the equations that governed the play of matter at the quantum level.

A major breakthrough came in 1927 when the British physicist Paul Dirac successfully blended quantum theory and special relativity in his theory of the electron. A big surprise of his work (not least to Dirac) was the prediction of the first antiparticle–a positively charged electron called the positron, which was found six years later by the American Carl Anderson. Armed with Dirac's electron theory, scientists hoped to inject quantum theory quickly into the whole of electromagnetism. But this didn't prove so straightforward. Whenever attempts were made to solve the quantized versions of Maxwell's equations, the results blew up in the faces of the frustrated theorists. Instead of getting sensible values for quantities such as the mass and charge of the electron, what popped up was the silly answer of infinity.

Finally, in the late 1940s, a trio of physicists–Richard Feynman at Cornell, Julian Schwinger at Harvard, and Sin-itiro Tomonaga in Japan–pulled a mathematical rabbit out of the hat called renormalization that saved the day. The "War of Infinities," as it became known, fought in the shadow of a far more momentous conflict, had been won. Its upshot was quantum electrodynamics (QED). So successful was this new theory that it enabled the properties of the electron, and its heavier cousin the muon, to be calculated correctly to an astonishing ten significant figures after the decimal point.

QED also gives us a physical insight into how a force is transmitted between two matter particles such as electrons. Newton had problems with the idea of action-at-a-distance, as did Maxwell. Quantum mechanics, however, profoundly alters classical field theory. In the classical scheme of things, energy and momentum are continuous quantities somehow (though one's never quite sure how) carried by the field. In quantum mechanics, however, energy and momentum are fragmented into tiny discrete units–quanta–which show up as particles. These field particles act as little messengers, conveying the force by traveling between the interacting particles of matter, just as a baseball pitcher transfers energy and momentum to the catcher when he hurls a ball. In QED, the messenger or "exchange" particles are none other than photons–particles of light. In other words, photons are the quanta of the electromagnetic field.

Each of the forces is mediated by its own kind of exchange particle, which has a definite rest mass (zero in the case of photons) and an intrinsic spin, or angular momentum, that can take integral values, such as 0, 1, or 2. Such particles are known as *bosons.* Particles of matter, on the other hand, have half-integral spin, such as ½ or 3/2, and belong to the family known as *fermions.* Bosons produce forces with distance ranges that vary as one divided by the particles' mass. Consequently, forces mediated by massive particles, such as the weak force, act over only a limited range, whereas forces mediated by massless particles, such as electromagnetism and gravity, have an apparently infinite range, with the force diminishing in strength inversely as the square of the distance between the interacting particles.

So amazingly effective did QED prove that researchers began looking for ways to extend and adapt its mathematical framework to the other forces in nature. A lot of effort went into trying to describe the weak force with equations similar in form to those of quantum electrodynamics. This led to the prediction of a new set of exchange particles, called intermediate bosons and first detected in the 1970s, which would carry the weak force. In 1967 the American physicists Steven Weinberg and Sheldon Glashow and the Pakistani physicist Abdus Salam went one better by showing how the weak force and electromagnetism could be combined to give something called the *electroweak interaction.* In today's universe the weak force and electromagnetism are completely separate and distinct. But for the first trillionth of a second or so after the Big Bang, conditions were so extreme that the two forces were united. During that initial sliver of cosmic time, the temperature was sufficiently high that the weak force and electromagnetism were one and the same. Thereafter, the symmetry that had made them identical was broken or, rather, hidden, as when a pond of water, in which there's a high degree of internal symmetry, freezes to ice in which the level of symmetry is less. Modern particle accelerators can reach energies comparable to those in the electroweak phase of the infant universe and have successfully tested theoretical predictions of this unified force.

Following the triumph of the electroweak synthesis, physicists were emboldened to try to corral the strong force with the electroweak in what became known as a grand unified theory, or GUT. The strong force operates between quarks–the fundamental building blocks of *hadrons*, or heavyweight particles, of which protons and neutrons are the most familiar examples. It turns out that a theory very much like QED, known as quantum chromodynamics (QCD), does a good job of describing strong interactions in terms of exchange particles called gluons. But attempts to wed QCD with the electroweak theory to yield a credible GUT haven't met with unqualified success. The trouble is that most of the GUTs that make sense also lead to some startling predictions, such as the decay of the proton (which had been thought utterly stable), solitary north and south poles known as *magnetic monopoles*, and a new class of particles called *leptobosons*. Unfortunately, despite determined efforts, not a scrap of evidence for these exotica has been found. Many researchers, therefore, are now focusing on a different and more ambitious approach. They're bypassing the GUT stage and trying to unify *all* the four basic forces–electromagnetism, the strong and weak forces, and gravity–in one fell swoop.

Gravity has been a real stumbling block in the quest for the grail of final unification. It is such a different beast from the other three kinds of interaction. Consider its strength, or lack of it: a pair of protons push apart more powerfully through electromagnetism (because of their like positive charges) than they pull together through gravitation by a staggering factor of 100 million trillion trillion trillion. Gravity–the planet builder, the apple dropper, the star crusher–is, in reality, almost unimaginably feeble. The only reason it's so influential on astronomical scales is that space, overall, is electrically neutral. The feebleness of gravity, compared with the other three fundamental forces, makes it impossible to study at the particle level in laboratories. Even an accelerator as big as a planet wouldn't bring the graininess of gravity into view.

Another problem is that existing quantum field theory depends on particle fields that never set foot outside the flat Minkowskian space-time of special relativity. General relativity, in complete contrast, depends crucially on modeling gravity as a curvature within space-time that changes as mass moves. Put another way, unlike in Newtonian mechanics and special relativity, there is no fixed space-time background in general relativity–the geometry of space-time is always on the move. The scenery and stage is constantly shifting, as are the actors. You might say that general relativity is a relational theory in which the only physically relevant information is the relationship between different events in space-time.

When theorists tried the most obvious ways of combining quantum field theory and general relativity–for example, by treating gravity as just another particle field–they quickly ran into the kind of problem with infinites that held up the early developers of QED, only on a much bigger scale. Because gravity particles attract one another, adding together all of their interactions results in a mountain of infinities that can't easily be canceled out to yield sensible, finite results. This is a different ballgame than in QED, where interactions that lead to infinite results are few enough in number to be removable by the trick of renormalization.

Despite these difficulties, theorists are confident they'll eventually be able to come up with a quantum theory of gravity. But they don't expect it will resemble anything like a stitching-together of current quantum field theory and general relativity.

Good Vibrations

One of the major outcomes of work on the electroweak unification was the so-called Standard Model, which neatly described all the elementary particles in nature and the forces between them–with one notable exception. It included six different types of lepton, or lightweight particle, six different types of quark, and the exchange particles for the weak, strong, and electromagnetic interactions. It also called upon an enigmatic new particle, named after its inventor, Peter Higgs of the University of Manchester (England), which, although not yet detected, is expected to play an important role in fixing the masses of all the other particles in the scheme. The Standard Model has agreed well, so far, with experimental data collected using particle accelerators. Yet physicists aren't completely happy with it. For one thing, it leaves too many arbitrary properties undecided–important values that simply have to be stuck into the model ad hoc. For another, it has no place for gravity.

How then to get gravity into the scheme? A clue to this emerged while researchers were working on the quantum field theory of the strong force. Along the way, they came up with a wonderfully creative explanation for the observed relationship between the mass and spin of hadrons. Called string theory, it treats particles as specific vibrations or excitations of very, very small lengths of a peculiar kind of string. In the end, QCD proved to be a better theory for hadrons. Yet string theory wasn't consigned to the trash can of ideas that had passed their sell-by date. It made one extremely interesting prediction: the existence of a particle–a certain excitation of string–with a rest mass of zero and an intrinsic spin of two units. Theorists had long known that there ought to be such a particle. It was none other than the hypothetical exchange particle of gravitation: the graviton.

With this discovery, that one of the essential vibrational modes of string corresponded to the graviton, string theorists realized they had a bigger fish to fry than trying to explain the ins and outs of hadrons. Their notions of elemental quivering threads might, it seemed, bear directly on the much sought-after quantum theory of gravity–and not just because the graviton is predicted by string theory. You can stick a graviton into quantum field theory by hand if you like, but it won't do you any good because you'll be blown away by infinities. *Particle* interactions happen at single points in space-time, so that the distance between interacting particles is zero. In the case of gravitons, the mathematics behaves so badly at zero distance that the answers come out as gobbledygook. String theory gets around this problem because the interacting entities aren't points but lengths that collide over a small but finite distance. As a result, the math doesn't self-destruct, and the answers make sense.

To get the hang of string theory, think of a guitar string that's been tuned by stretching it between the head and the bridge. Depending on how the string is plucked and how tense it is, different musical notes are created. These notes can be thought of as excitation modes of the guitar string under tension. Similarly, in string theory, the elementary particles observed in particle accelerators correspond to the notes or excitation modes of elementary strings. One mode of vibration makes the string appear as an electron, another as a photon, and so on.

In string theory, as in guitar playing, the string has to be under tension in order to become excited. A big difference is that the strings in string theory aren't tied down to anything, but instead are floating in space-time. Even so, they're under tension–by an amount that depends, roughly speaking, on one over the square of the string's length. Now, if string theory is to work as a theory of quantum gravity, then the average length of a string has to be approximately the distance over which the quantization of space-time–the granularity of space and time–becomes noticeable. This outrageously tiny distance, known as the Planck length, is about 10^{-33} centimeters, or one billion trillion trillionth of a centimeter. So much tinier is it than anything that current or planned particle physics technology can hope to be able to see that string theorists have to look for craftier, more indirect ways to test their ideas.

String theories come in various forms. All of these assume that the basic stuff of creation are tiny wriggling strings. However, if the theory deals with only *closed* loops of string, like Spaghetti-Os, then it's limited to describing bosons–the force-carrying particles–and so is called bosonic string theory. The first string theory to be developed was of this type. If open strings, like strands of ordinary spaghetti, are allowed into the theoretical picture, they provide a description of fermions, or particles of matter. But a very interesting thing happens when string theory is extended in this way to let in fermions. It demands that there must be a special kind of symmetry in the particle world known as supersymmetry. In this expanded master plan of things, there's a corresponding fermion for every boson. In other words, supersymmetry relates the particles that transmit forces to the particles that make up matter.

A supersymmetric string theory is called, not surprisingly, a superstring theory.

Theorists uncovered three different string theories that were mathematically consistent and therefore made good sense. Two of these were bosonic; the other was of the superstring ilk. But in order to make any of them work, they had to resort to a strategy first employed by Kaluza and Klein in the days when Einstein first started wandering down his blind unification alley: they had to call upon higher dimensions, rolled up so small that they were way below the threshold of detection. The bosonic string theories needed an awesome twenty-six dimensions (twenty-five of space plus one of time) in order to work properly, which seemed a bit of a stretch even for scientists who enjoyed some way-out sci-fi in their off hours. Compared with this, the mere ten dimensions of space-time required by superstring theory seemed positively modest. Six of the ten would have to be curled up, or "compactified," to leave visible the four normal space-time dimensions (three of space plus one of time). But these compactified dimensions, far from being an embarrassment to be swept under the cosmic carpet and forgotten about, would come in very handy if string theory were to aspire to become a theory of everything; motion in them could be used to explain the values taken by important constants in nature, such as the charge on the electron.

Combining the best features of bosonic and superstring theory has led to two other consistent schemes known as heterotic string theories. There are now five viable string theories in all, which, if we're hoping to arrive at the one true TOE, is a tad too many. Fortunately, it's beginning to look as if the quintet of finalists for the Miss Universe Theory competition is really the same contestant dressed up in five different costumes. This supersymmetric mistress of disguise has been given the rather enigmatic name M-theory.

Some say that the M is for Mother of All Theories. Others say that it stands for Magic or Mystery. But, although no one seems to know for sure, there may be a more prosaic reason for this particular choice of initial.

Before string theory rose to scientific superstardom, the most popular unified theory in town was *supergravity*, which was basically supersymmetry plus gravity without the string. Like any respectable quantum gravity candidate, it boasted a surfeit of space-time dimensions–in this case, eleven (the compactified ones all wrapped up neatly on an itty-bitty, seven-dimensional sphere). Unfortunately, it had to be abandoned because of the problems mentioned earlier involving point particles and string.

Along came M-theory. Still under development, it carries the hopes of many that it will combine the various flavors of string theory soup into one single, satisfying broth. The cost of this in conceptual terms is the addition of a single dimension: M-theory is eleven-dimensional but has the unusual trait that it can appear 10-dimensional at some points in its space of parameters. Supergravity rides again–but this time with strings attached.

And the M in M-theory? We omitted to say earlier that while strings, with their one-dimensional extension, are the fundamental objects in string theory, they are not the only objects allowed. String theory can accommodate multidimensional entities, called branes, with anywhere from zero (points) to nine spatial dimensions. A brane with an unspecified number, p, of dimensions is called a p-brane. In M-theory, with its extra dimension, the fundamental object is an M-brane, which resembles a sheet or membrane. Like a drinking straw seen at a distance, the membranes would look like strings since the eleventh dimension is compactified into a small circle. Membranes, M-branes, M-theory. Hmmm . . .

An End in Sight?

Building a theory of everything is one thing; testing it is quite another. The physical conditions that have to prevail for the four forces of nature to be unified into a single force haven't existed since the universe was about 10^{-43} seconds (one ten million trillion trillion trillionth of a second) old. There is not the remotest chance of re-creating that kind of environment in the lab anytime soon, if ever. But what physicists can do is look for other clues that their unification scheme is on the right track.

We saw earlier that supersymmetry predicts that there are supersymmetric fermion partners of all the force-carrying bosons. The supersymmetric partner of the graviton, for example, is a spin $^3/_2$ particle that, like all its supersymmetry cousins, is expected to be very massive—maybe a thousand times more massive than a proton. This high mass has put the creation of such particles beyond the reach of accelerators thus far. But that may be about to change. A new generation of more powerful instruments, including the Large Hadron Collider (LHC) at the European CERN (Centre Européen pour la Recherche Nucleaire) facility near Geneva, Switzerland, is about to come on line and be capable of exploring the energy domain in which the new particles might be found. Evidence for supersymmetry at high energy would be compelling evidence that string theory was a good mathematical model for nature at the smallest distance scales.

In some ways, the invention of string theory was premature—its physical concepts running ahead of the mathematical techniques needed to describe them. One of the architects of string theory in its modern form, Edward Witten of the Institute for Advanced Study in Princeton (where Einstein spent his latter days), has said:

> By rights, twentieth century physicists shouldn't have had the privilege of studying this theory. What should have happened, by rights, is that the correct mathematical structures should have been developed in the twenty-first or twenty-second century, and then finally physicists should have invented string theory as a physical theory that is made possible by those structures. . . . [T]hen the first physicists working with string theory would have known what they were doing, just like Einstein knew what he was doing when he invented general relativity.

There are other theories of quantum gravity besides string theory. One of the leading rivals is called loop quantum gravity, founded in the late 1980s by Abhay Ashtekar of Penn State University, Carlo Rovelli of the Center for Theoretical Physics in Marseille, France, and Lee Smolin of Harvard. Its strategy is to focus on quantizing the space-time of general relativity without getting involved in trying to unify gravity with the three other forces. Smolin, however, has suggested that string theory and loop quantum gravity might eventually be reconciled as different aspects of the same underlying theory.

Thales, Anaximenes, Newton, Einstein–all sought the holy grail of unification. They would have been pleased to know their descendants were now perhaps closing in on the great prize and what it might make possible.

13

Engineers of the Continuum

Listen; there's a hell of a good universe next door: let's go.

—E. E. CUMMINGS

"Knowledge is power" the saying goes. Certainly, understanding a force is often the prelude to controlling it. Our ability to produce, distribute, and harness electricity stems from a sound knowledge of electromagnetic theory. Our ability to land on the moon or fly space probes on accurate paths to the planets rests on a thorough understanding of Newtonian gravity, which is perfectly good enough for interplanetary jaunts. As we learn more about the nature of gravity, especially at the quantum level, we can expect our powers over gravity to grow. And if in the future we can learn to control or modify the very architecture of space-time, the implications for space travel and for a new science–experimental cosmology–will be quite extraordinary.

A Shortcut to Anywhen

Interest in black holes and wormholes has led people to speculate that these extreme distortions in the fabric of space-time might someday be used for quick access to the stars and even as a way of traveling through time. At a superficial level, the idea is easy to grasp and attractive. One way to think of a wormhole is as a shortcut between different parts of our universe or as a connecting tube with other universes. Pop down one of its openings, zip along its relatively short length, and you could reappear moments or minutes later (as experienced by you) at the other end, many light-years away. Anyone who's seen episodes of *Stargate SG-1* or the Star Trek series *Deep Space Nine* (*DS9*) will be familiar with the concept. In the case of *DS9*, a stable wormhole is found that links opposite sides of the galaxy by a journey that's as convenient as a short taxi ride.

All we need do, it seems, is find a handy wormhole or two, and we've got intragalactic, and maybe even intergalactic or intracosmic, travel licked. But there may be a few snafus along the way. Problem 1: We don't even know if natural wormholes exist. Problem 2: If they do, we've no idea where they are and where they lead (if anywhere). Problem 3: You'd have to be a bit concerned about what would happen to any astronaut who ventured into one; wormhole travel is not for the fainthearted.

When Carl Sagan was writing his novel *Contact*[32] about a SETI researcher who discovers a message coming from the direction of the star Vega, he wanted to be able to whisk his fictional scientist to that remote location to meet the aliens in person. So he called his buddy Kip Thorne, a top theorist in gravitational physics at Caltech, and asked for his advice. "I want the science to be as accurate as possible," said Sagan. Was it theoretically feasible, he wondered, to send a person in some kind of a vessel through a conduit in space-time? Thorne asked two of his grad students, Michael Morris and Uri

Yurtsever, to look into the problem. The upshot of their work was a paper[23] published in 1988 that gives a provisional thumbs-up to wormhole transportation; it's also possibly the first paper in modern times to be written directly as a result of a science-fiction story.

Morris, Thorne, and Yurtsever realized that a journey such as the type Sagan had in mind would be successful only if the entrance of the space-time passage could be relied on to stay open. Theory warns that one of the potential showstoppers to using wormholes is that like old mine workings held up by rickety posts, they're prone to collapse. Even the slightest bit of matter entering the wormhole's mouth is liable to make the whole thing implode. The Caltech trio figured that to keep a wormhole propped open you would need a pretty unusual means of support. It would involve setting up a region that had a *negative* energy density–a negative amount of energy per unit volume–which is a situation you hardly ever come across in the real world. One of the few places negative energy density occurs on Earth is in the gap between two very close parallel metal plates. The fact that there's more energy in the space around them than between them means that the plates feel an ever-so-slight force pushing them together. Because this effect was discovered in 1948 by the Dutch physicist Hendrick Casimir, it has been named after him.

Of course, it would be very handy if nature provided us with ready-made wormholes. That isn't as fantastic as it sounds because the original expansion of the universe was so extraordinarily fast and explosive that even tiny wormholes spawned during the Big Bang might have been stretched and blown up to macroscopic size before they had time to disappear. But the chances of any of these ancient relics connecting places we'd want to travel from or to are pretty small. It would be like stumbling across tunnels and shafts leading to all the world's richest underground deposits of gold, diamonds, and other treasures. And even if, by chance, a space-time tunnel were fortuitously placed, it would most likely be unstable and therefore useless. The odds are that any wormholes we need are going to have to be made and maintained by ourselves. The question is, how?

For an answer, Thorne and his students turned their attention to empty space. Even the best vacuum isn't really empty but instead, on the smallest of distance scales, comparable with the Planck length, seethes with violent fluctuations in the geometry of space-time. At this level of nature, known as the quantum vacuum, fantastically small wormholes are thought to continuously wink into and out of existence like the kinetic patterns made by froth and foam on the sea. The Caltech researchers suggested that at some point in the future we might learn how to surgically remove one of these tiny wormholes, isolate it, and inflate it to macroscopic size by adding energy.

The wormhole would then be stabilized with a region of negative energy density, for which the Caltech team suggested an ultrapowerful version of the Casimir effect apparatus. Two charged superconducting spheres would be placed in the wormhole mouths, which, to begin with, might be very close together. These mouths would then be positioned where they were needed. For example, one of them might be set up at the edge of the solar system while the other was loaded aboard a spaceship and flown to some location many light-years away. Because the delivery trip would be through normal space-time, it would have to take place at sublight speed. But both during the trip and afterward, instantaneous communication and transport through the wormhole would be possible. The ship could even be supplied with fuel and provisions through the mouth it was carrying. And thanks to the time-dilation effect of special relativity, the journey wouldn't have to take long, even as measured by Earth-based observers. For example, if a fast starship carrying a wormhole mouth were to travel to Vega, twenty-five light-years away, at 99.995 percent of the speed of light (giving a time-dilation factor of 100), shipboard clocks would measure the journey as taking just three months. But the wormhole stretching from the ship to Earth would directly link the space and time between both mouths–the one on the ship and the one left behind on (or near) Earth. Therefore, as measured by earthbound clocks, too, the trip would have taken only three months–three months to establish a more-or-less instantaneous transport and communications link between here and Vega.

All this sounds wonderful, but there's a troublesome aspect to the Thorne-Morris-Yurtsever type of wormhole, which has to do with the way it's stabilized. The Casimir-effect spheres that hold the mouths open would generate such incredibly powerful forces that they would probably tear apart anything or anyone that tried to pass through. In an effort to design a more benign environment for wormhole travelers, Matt Visser of Washington University in St. Louis conceived an arrangement in which the space-time aperture of a wormhole mouth is kept essentially flat and therefore force-free. The trick he came up with was to frame the mouth with struts of exotic matter–weird, hypothetical stuff that has an extremely high negative energy density and is therefore repelled rather than attracted by normal gravity. Visser envisaged a cubic design, with flat-space wormhole connections on the square sides and cosmic strings as the edges. Each cube-face might connect to the face of another wormhole-mouth cube, or the six cube faces might connect to six different cube faces in six separated locations.

Even the best theories warn that we face major engineering problems in building a workable wormhole subway. Either the wormholes end up incredibly narrow, with a throat radius you couldn't squeeze a single atom through, or if they're macroscopic, the negative energy needed to keep them open has to be confined to incredibly thin bands. For example, according to one study, a wormhole with a throat radius of one meter would need to be supported by negative energy in a band no thicker than a billion trillionth of meter–a millionth the size of a proton. Matt Visser has estimated that the negative energy required for this size of wormhole is equal in size to the total energy generated by 10 billion stars in one year. The situation doesn't get much better for larger wormholes. The maximum allowed thickness of the negative energy band turns out to be proportional to the cube root of the throat radius. Even if the throat radius is increased to a size of one light-year, the negative energy still has to be confined to a region smaller than a proton radius and the total amount required increases linearly with the throat size.

Wormhole engineers clearly confront daunting problems. They have to find a way to confine large amounts of negative energy within extremely small volumes. So-called cosmic strings, hypothesized in some cosmological theories, involve immense energy densities in long, narrow lines. All known physically reasonable cosmic-string models, however, have positive energy densities.

Given that our technology isn't yet up to the task of building a wormhole subway, the question arises as to whether one might already exist. One possibility is that advanced civilizations elsewhere in the galaxy or beyond have already set up a network of wormholes that we could learn to use. This was the premise used by Sagan in his novel that enabled his heroine, Ellie Arroway, to hop to Vega and beyond.

Another controversial idea is that wormholes might not be restricted to high-speed space jaunts. They could conceivably also work as time machines. Not everyone is happy with this idea, however, because it might lead to all sorts of bizarre paradoxes. Stephen Hawking was so concerned about the problems that time hopping might create that in the 1980s he argued that something fundamental in the laws of physics conspires to prevent wormholes from being exploited for time travel. This idea forms the basis of what's called Hawking's Chronology Protection conjecture.

Danger: Black Holes on the Loose

Of more immediate interest–and concern–is the possibility that we might be able to fabricate our own black holes. The amount of energy that can be concentrated in one place by modern particle accelerators or high-power lasers is phenomenal. It might even be enough to cause a tiny region of space-time, smaller than a pinprick, to collapse in upon itself and disappear inside its own event horizon.

In 2005 a report surfaced from the Relativistic Heavy Ion Collider (RHIC), an accelerator laboratory in New York, that events resembling this had already taken place. At the RHIC beams of gold nuclei are smashed together at near light speeds, causing the nuclei to be broken down into their constituent quarks and gluons, which then form balls of plasma three hundred times hotter than the surface of the sun. Although the fireballs last a mere 10 million billion billionths of a second, they can be detected because they absorb jets of particles produced by the beam collisions.

When Horatiu Nastase, a researcher at Brown University, analyzed what was happening inside the fireballs, he found something odd: ten times as many jets were being absorbed as were predicted by calculations. Particles disappear into the fireball's core, Nastase believes, and reappear as thermal radiation, just as matter that falls into a black hole will eventually be reemitted as Hawking radiation. The RHIC may be producing only analogs of black holes, but that still makes those analogs a useful tool for studying some of the phenomena associated with the real thing. For instance, the rapid slowing down of RHIC nuclei as they smash into each other very briefly, in less than a billion trillionth of a second, is similar to the extreme gravitational environment close to a black hole.

Soon we may not need to rely on simulations. Scientists seriously expect that the next generation of particle accelerators, spearheaded by CERN's Large Hadron Collider (LHC), will spawn microscopic black holes on a regular basis. The LHC will bring protons and antiprotons together with such a force that the collision will create temperatures and energy densities not seen since the first trillionth of a second after the Big Bang. This should be enough to pop off numerous tiny black holes–up to one a second–with masses of just a few hundred protons. Another more controlled way of manufacturing small black holes may be the focusing of ultrapowerful laser beams at a single point in space. Any fears that these miniature gravitational monsters might break loose and start gobbling up the planet can safely be set aside. Similar events probably routinely take place high in the Earth's atmosphere as high-energy cosmic rays from space smash into air molecules. Black holes that are this tiny will evaporate almost instantly, their existence detectable only by dying bursts of Hawking radiation. Those death flashes, if witnessed in the lab, are what researchers hope will give us a deeper insight into black hole physics.

It's still a mystery whether Hawking radiation contains any useful information about the particles that formed the black hole in the first place or fell into it later. The particles at the time of their disappearance had charge, spin, and other fundamental characteristics that may not have been erased by the black hole.

Also, the exact manner in which a black hole expires may give us a view into the higher, compactified dimensions in space.

Playing God

Being able to create space-time artifacts on demand may, in the long term, open up an even more fantastic possibility for science: the fabrication of new, experimental universes. This bizarre, seemingly outrageous notion follows straight from some of the work that's been done in trying to refine the details of the Big Bang in which our own universe was created. One of the more extraordinary consequences of the inflation scenario is that a universe can be formed essentially from nothing. Cosmogenesis, it seems, is triggered when a certain region of space-time enters a state called the false vacuum, which then destabilizes and rapidly expands like a weak spot on a balloon that suddenly blows out when being inflated. Because the positive energy of matter effectively cancels out the negative energy of gravity, creation turns out to be cheap; the universe may, as MIT physicist Alan Guth put it, be a free lunch.

Another of the architects of inflation theory, Andre Linde, at Stanford University, believes these ideas may open the door to the synthesis of universes in the lab. His version of inflation, known as chaotic inflation theory, sees universes being created naturally all the time from a kind of primeval ocean of foaming, frothing space-time. To get a universe like ours started, he found, takes only a hundred-thousandth of a gram of matter. That's enough to generate a small chunk of vacuum that blows up into the billions and billions of galaxies we see around us. It seems like cheating, but this is how inflation theory works: all the matter in the universe gets created from the negative energy of the gravitational field. So, asks Linde, if it's that easy, what's to stop us from cooking up our own free cosmic lunch in the laboratory kitchen?

It might seem as if there's a potentially disastrous drawback to this ambitious scheme. If we triggered a mini-Big Bang here and now, wouldn't the baby universe we fashioned expand into our own universe, destroying Earth and eventually pushing aside everything else like a cosmic airbag going off? Linde thinks not. The new universe, he explains, would expand *into itself*, not into any preexisting region of space-time. Its space would be so curved that it would appear to us as tiny as an elementary particle. In fact, it might end up disappearing altogether from the world of its creators.

Taking these fantastic ideas further, Linde asks whether it might be possible for the makers of a new universe to keep track of their progeny. Could we continue to study the results of our genesis-in-a-bottle–maybe even communicate with any creatures that came to inhabit it and perhaps offer them some of the benefits of our own scientific and technological wisdom? One of the problems with getting a message into a baby universe is that the babyverse would blow up very quickly because of inflation and thereafter exist as its own bubble of space-time. Its entire history might run its course in, what to us, seemed the wink of an eye. Likewise, from the point of view of someone in the baby universe, our own cosmic story might appear to last only an instant.

Even so, thought Linde, there might be a way we could let our creations know where they'd come from or, at least, that they lived in a very special, designer cosmos. By manipulating the cosmic seed in the right way, we would have the power to ordain certain physical parameters of any universe we brought into being. The so-called initial conditions are up for grabs in natural cosmogenesis. By fabricating universes ourselves, however, we could, in principle, fine-tune these parameters at the outset. For example, we might fix a particular value for the ratio of the electron's mass to that of the proton. Such ratios–constants of nature–look like arbitrary numbers to us, but there is no obvious reason they should take one value rather than another. If we assumed the role of creator, we could, by fixing certain values for constants, encode a subtle message into the very structure of the universe. Of course, this message would be legible only to physicists who came to inhabit the manufactured cosmos. Perhaps it's for the best that the denizens of the new universe could only become aware of the true nature of their circumstances when their science and technology had reached a certain level of sophistication.

We're led by such thinking to a disquieting possibility: perhaps all this has already happened. Some scientists have pointed out how curiously contrived are the circumstances of our own existence. Had the relative strengths of the four basic forces, for example, been even slightly different, it's unlikely that galaxies, stars, and life could have come about. Do we ourselves live in a designer universe? And if so, is that because, of all the multiplicity of universes that can be spawned naturally in Linde's chaotic inflationary theory, only those of a certain type are suitable for habitation? Or do the special fixings and tunings point to an intelligent creative hand at work?

Science becomes enmeshed with theology at this point. Those who favor a religious perspective might argue that God would, of course, have to set the initial parameters so that the universe would be conducive to life. But Linde's notion of cosmos creation in a lab might suggest to others that we inhabit a universe that is the product of some other civilization's scientific experiments. Even if our gods turned out to be super-advanced aliens in another universe, the theologian can still ask where those aliens came from. Another race in another universe perhaps? The buck has to stop somewhere.

The Big Escape

Everything comes to an end eventually: books, governments, lives, stars, universes. One way or another, our cosmos is doomed. Present theory and observations suggest the universe is facing runaway acceleration. The far future will bring increasing isolation of galaxy clusters, and in the depths of time to come, a big freeze will occur when all remaining useful energy reserves have been exploited. The philosopher and mathematician Bertrand Russell spoke of the "unyielding despair" he felt when contemplating what lies ahead: "All the labors of the ages, all the devotion, all the inspiration, all the noonday brightness of human genius, are destined to extinction in the vast death of the solar system; and the whole temple of man's achievement must inevitably be buried beneath the debris of a universe in ruins."

At some remote date, the last star will wink out, and all the universe will have left to speak of is a far-flung scattering of dead suns, cooling neutron stars, and black holes. Any remaining civilizations would presumably gather around the last flickering embers of this nuclear debris, finally clustering like homeless refugees around black holes emitting a faint glow of Hawking radiation.

The really smart races, however, won't wait that long. They will make good their escape from the done-for cosmos. Just as, at some future date, the offspring of humanity will doubtless migrate to other stars and galaxies, so, in deep time, the great races and minds of the universe will pack up their bags and leave–for a new space-time continuum. After all, if we can learn the trick of genesis in our labs or create wormholes to other places and times, or perhaps even establish communication with some of the countless inhabited universes that may comprise the great multiverse predicted by chaotic inflationary theory, then we, or our descendants, can surely figure out how to travel in person to an alternate reality.

It's a rule of evolution that when the environment changes radically, life must confront the choice of dying, adapting, or fleeing. Since death isn't an acceptable option to a brainy species, and limitless adaptation to a universe that's continually running down isn't feasible, departure becomes, in the end, a necessity. Fortunately, we have many billions, if not trillions, of years to prepare our plans for the great escape. A first step–one that, even now, we've embarked upon–will be to gain a thorough understanding of how gravity works at the quantum level. This knowledge will ultimately prove crucial in calculating factors such as the stability of wormholes that may help connect us to a parallel universe and in establishing, in advance, the nature of the cosmos that lies on the other side of the space-time umbilical.

Gravity, as its name implies, is serious stuff. For now, it helps us keep our feet on the ground. But in the remotest of futures, it will be the key to our continued survival and our exploration of the endless stretches of space and time.

References

1. Allais, M. Should the Laws of Gravitation Be Reconsidered? *Aero-Space Engineering* 9 (1959): 46–52.
2. Anderson, J. D., M. M. Nieto, and S. G. Turyshev: A Mission to Test the Pioneer Anomaly. *International Journal of Modern Physics* D 11(2002): 1545–51.
3. Anderson, J. D., P. A. Laing, E. L. Lau, A. S. Liu, M. M. Nieto, and S. G. Turyshev. Study of the Anomalous Acceleration of Pioneer 10 and 11. *Physical Review* D 65 (2002): 082004.
4. Bekenstein, J. D. Relativistic Gravitation Theory for the MOND Paradigm. In *Second Canadian Conference on General Relativity and Relativistic Astrophysics*, ed. A. Coley, C. Dyer, and T. Tupper, 68. Singapore: World Scientific, 1987.
5. Chandrasekhar, S. The Maximum Mass of Ideal White Dwarfs. *Astrophysical Journal* 74 (1931): 81–82.
6. Duif, Chris P. A Review of Conventional Explanations of Anomalous Observations During Solar Eclipses. arXiv gr-qc/040802 v3, last accessed October 8, 2004.
7. Dyson, F. W., A. S. Eddington, and C. Davidson. *Philosophical Transactions of the Royal Society of London* 220A (1920): 291–333.
8. Einstein, A. Zur Elektrodynamik bewegter Körper (On the Electrodynamics of Moving Bodies). *Annalen der Physik* 17 (1905): 891–921.
9. Einstein, A. Ist die Trägheit eines Körpers von seinem Energieinhalt Abhängig? (Does the Inertia of a Body Depend upon Its Energy Content?). *Annalen der Physik* 18 (1905): 639–41.
10. Einstein, A. Feldgleichungen der Gravitation (The Field Equations of Gravitation). *Sitzungsberichte der Preussischen Akademie der Wissenschaften* (1915).
11. Einstein, A. Kosmologische Betractungen zur allgemeinen Relativitätstheorie. *Sitzungsbericte der Preussischen Akadamie der Wissenschaften*

(1917). Reprinted as "Cosmological Considerations on the General Theory of Relativity." In *Principles of Relativity*, trans. W. Perrett and G. B. Jeffery. New York: Dover, 1923.

12. Einstein, A., and M. Grossman. Entwurf einer Verallgemeinerten Relativitätstheorie und eine Theorie der Gravitation (Draft of a Generalized Theory of Relativity and a Theory of Gravitation). I. Physikalishcher Teil von (Physics Part by) A. Einstein. II. Mathematischer Teil von (Mathematics Part by) M. Grossman. *Zeitschrift für Physik* 62 (1913).
13. Fitzgerald, G. The Ether and the Earth's Atmosphere. *Science* 13 (1889): 390.
14. Hilbert, D. Die Grundlagen der Physik (Erste Mitteilung) (The Foundations of Physics, First Communication). *Nachrichten von der Gesellschaft der Wissenschaften zu Gottingen, Mathematisch-physikalische Klasse* (1915): 395–407.
15. Hulse, R. A., and J. H. Taylor. Discovery of a Pulsar in a Binary System. *Astrophysical Journal* 195 (1975): L51–L53.
16. Jeverdan, G. T., G. I. Rusu, and V. Antonescu. Experiments Using the Foucault Pendulum during the Solar Eclipse of 15 February 1961. *Biblical Astronomer* 1 (1981): 18–20.
17. Kerr, R. P. Gravitational Field of a Spinning Mass as an Example of Algebraically Special Metrics. *Physical Review Letters* 11 (1963): 237–38.
18. Liais, Emmanuel. *Popular Science* 13 (1878): 732–35.
19. Luminet, J.-P., J. R. Weeks, A. Riazuelo, R. Lehoucq, and J.-P. Uzan. Dodecahedral Space Topology as an Explanation for Weak Wide-Angle Temperature Correlations in the Cosmic Microwave Background. *Nature* 425 (2003): 593–95.
20. Michell, John. On the Means of Discovering the Distance, Magnitude, etc. of the Fixed Stars. *Philosophical Transactions of the Royal Society of London* 74 (1784): 35–57.
21. Milgrom, M. MOND–Theoretical Aspects. *New Astronomy Review* 46 (2002): 741–53.
22. Mishra, D. C., and M. B. S. Rao. Temporal Variations in Gravity Field during Solar Eclipse on 24 October. *Current Science* 72 (1997): 783.
23. Morris, M. S., K. S. Thorne, and U. Yurtsever. Wormholes, Time Machines, and the Weak Energy Condition. *Physical Review Letters* 61 (1988): 1446–49.
24. Noordung, Hermann. *Das Problem der Befahrung des Weltraums* (The problem of spaceflight). Berlin: Schmidt and Co., 1928.
25. Oberth, Hermann. *Die Rakete zu den Planetenräumen* (The rocket into interplanetary space). Munich: R. Oldenbourg, 1923.

26. Oppenheimer, J. R., and G. M. Volkoff. On Massive Neutron Cores. *Physical Review* 55 (1939): 374–81.
27. Ott, E., C. Grebogi, and G. A. Yorke. Controlling Chaos. *Physical Review Letter* 64 (1990): 1196–99.
28. Page, G. L., D. S. Dixon, and J. F. Wallin. Utilizing Minor Planets to Assess the Gravitational Field in the Outer Solar System. arXiv: astro-ph/0504367, last accessed April 2005.
29. Perlmutter, S., G. Aldering, R. A. Goldhaber, P. Knop, P. G. Nugent, S. Castro, et al. Discovery of a Supernova Explosion at Half the Age of the Universe and Its Cosmological Implications. *Nature* 391 (1998): 51.
30. Poincaré, H. La Mesure du Temps (The Measurement of Time). *Revue de Métaphysique et de Morale* 6 (1898): 371–84.
31. Riess, A. G., et al. Observational Evidence from Supernovae for an Accelerating Universe and a Cosmological Constant. *The Astronomical Journal* 116 (1998): 1009–38.
32. Sagan, Carl. *Contact.* New York: Simon & Schuster, 1985.
33. Saxl, E., and M. Allen. 1970 Solar Eclipse as "Seen" by a Torsion Pendulum. *Physical Review* D 3 (1971): 823.
34. Schwarzschild, K. Ueber das Zulaessige Kruemmungsmaass des Raumes (On the Permissible Curvature of Space). *Vierteljahrsschrift der Ast. Ges.* 35 (1900): 337. Translated in J. M. Stewart and M. E. Stewart, On the Permissible Curvature of Space. *Classical and Quantum Gravity* 15 (1998): 2539–44.
35. Snow, C. P. *Variety of Men*, 85–86. Harmondsworth, U.K.: Penguin Books, 1969.
36. Taylor, J. H., L. A. Fowler, and J. M. Weisberg. Measurements of General Relativistic Effects in the Binary Pulsar PSR1913+16. *Nature* 277 (1979): 437.

Bibliography

Caspar, Max. *Kepler*. New York: Dover, 1993 (originally published 1948).

Chandrasekhar, S. *Newton's Principia for the Common Reader*. Oxford: Oxford University Press, 2003.

Einstein, Albert. *Relativity: The Special and the General Theory*. New York: Three Rivers Press, 1995.

Ferguson, Kittie. *Tycho and Kepler*. New York: Walker & Co., 2004.

Gamow, George. *Gravity*. New York: Dover, 2003 (originally published 1962).

Greene, Brian. *The Elegant Universe: Superstrings, Hidden Dimensions, and the Quest for the Ultimate Theory*. New York: W. W. Norton, 1999.

Greene, Brian. *The Fabric of the Cosmos: Space, Time, and the Texture of Reality*. New York: Knopf, 2004.

Halpern, Paul. *The Great Beyond: Higher Dimensions, Parallel Universes and the Extraordinary Search for a Theory of Everything*. Hoboken, N.J.: Wiley, 2004.

Kaku, Michio. *Parallel Worlds: A Journey through Creation, Higher Dimensions, and the Future of the Cosmos*. New York: Doubleday, 2004.

Schutz, Bernard. *Gravity from the Ground Up: An Introductory Guide to Gravity and General Relativity*. Cambridge: Cambridge University Press, 2003.

Shea, William R., and Mariano Artigas. *Galileo in Rome: The Rise and Fall of a Troublesome Genius*. Oxford: Oxford University Press, 2003.

Smolin, Lee. *Three Roads to Quantum Gravity*. New York: Basic Books, 2001.

Thorne, Kip S., Charles W. Misner, and John Archibald Wheeler. *Gravitation*. New York: W. H. Freeman, 1973.

Westfall, Richard S. *Never at Rest: A Biography of Isaac Newton*. Cambridge: Cambridge University Press, 1983.

White, Michael. *Isaac Newton: The Last Sorcerer*. New York: Perseus, 1998.

Index